Trade, Environment and Sustainable Development

A South Asian Perspective

Edited by

Veena Jha

Grant Hewison

and

Maree Underhill

Foreword by

Rubens Ricupero

Secretary General of UNCTAD

in association with
UNITED NATIONS CONFERENCE ON TRADE AND DEVELOPMENT
UNITED NATIONS DEVELOPMENT PROGRAMME

First published in Great Britain 1997 by

MACMILLAN PRESS LTD
Houndmills, Basingstoke, Hampshire RG21 6XS
and London
Companies and representatives
throughout the world

A catalogue record for this book is available
from the British Library.

ISBN 0-333-65134-0

First published in the United States of America 1997 by

ST. MARTIN'S PRESS, INC.,
Scholarly and Reference Division,
175 Fifth Avenue,
New York, N.Y. 10010

ISBN 0-312-16022-4

Library of Congress Cataloging-in-Publication Data
Trade, environment and sustainable development : a South Asian perspective / edited by Veena Jha, Grant Hewison, and Maree Underhill ; foreword by Rubens Ricupero.
p. cm.
Papers presented at a South Asian regional workshop and a national seminar on trade and the environment between Jan. 27–29, 1994.
"In association with United Nations Conference on Trade and Development, United Nations Development Programme."
Includes bibliographical references and index.
ISBN 0-312-16022-4
1. South Asia—Commerce—Environmental aspects—Congresses. 2. Environmental policy—South Asia—Congresses. 3. Sustainable development—South Asia—Congresses. 4. South Asia—Economic policy—Environmental aspects—Congresses. I. Jha, Veena, 1959– . II. Hewison, Grant. III. Underhill, Maree. IV. United Nations Conference on Trade and Development. V. United Nations Development Programme.
HF3770.3.A46T73 1997
338.954—DC20 96-11318
CIP

This book is printed on paper suitable for recycling and made from fully managed and sustained forest sources.

10 9 8 7 6 5 4 3 2 1
06 05 04 03 02 01 00 99 98 97

Printed in Great Britain by
The Ipswich Book Company Ltd
Ipswich, Suffolk

TRADE, ENVIRONMENT AND SUSTAINABLE DEVELOPMENT

Contents

Foreword by Rubens Ricupero vii

Notes on the Contributors ix

Introduction 1

PART I THE ISSUES

1 Trade, Environment and Sustainable Development
Kamal Nath 15

2 Environmentally Based Process and Production Method Standards: Some Implications for Developing Countries
René Vossenaar and Veena Jha 21

3 Environmentally Orientated Product Policies, Competitiveness and Market Access
Veena Jha and René Vossenaar 41

4 Environmental Policy Making, Eco-Labelling and Eco-Packaging in Germany and its Impact on Developing Countries
Christine Wyatt 51

5 Environmental Standards: Relocation of Production to the SAARC Region
Roland Mollerus 69

PART II COUNTRY CASE-STUDIES

6 Environmental and Trade Considerations in the South Asian Region
Prodipto Ghosh, Preeti Soni, M. C. Verma and Rakesh Shahani 81

7 The Sustainable Development of Leather Industries in Bangladesh
Fasih Uddin Mahtab 97

8 Trade and the Environment: A Perspective from Bhutan
Achyut Bhandari 105

9 Making Trade and Environmental Policies Compatible: Considerations from India
R. C. Jhamtani 109

10 Protection of the Environment, Trade and India's Leather Exports
Ashok Jha 117

11 The Impact of Environmental Standards and Regulations Set in Foreign Markets on India's Exports
Vasantha Bharucha 123

12 Making Trade and Environmental Policy Making Mutually Compatible in Nepal
Khilendra N. Rana 143

13 Making Trade and Environmental Policy Making Mutually Compatible in Pakistan
Akhtar Hasan Khan 149

14 Trade and the Environment: A Case Study from Sri Lanka
Lalith Heengama 155

15 Trade and the Environment: A Sri Lankan Perspective
V. Kanesalingam 161

PART III PATHS FORWARD

16 Trade, Environment and the Transfer of Environmentally Sound Technology
Veena Jha and Ana Paola Teixeira 171

17 The Transfer of Environmentally Sound Technology with Special Reference to India
Amrita N. Achanta, Pradeep Dadhich, Prodipto Ghosh and Ligia Noronha 185

18 Trade, Environment and Development Cooperation
Ebba Dohlman 203

19 Principles for Making Trade and the Environment Mutually Compatible
Nevin Shaw 209

20 Conclusions and Policy Recommendations
Veena Jha 217

Index 221

Foreword

In January 1994, UNCTAD and the Tata Energy Research Institute of India (TERI) jointly organised a South Asian regional workshop and an Indian national seminar on Trade and the Environment in New Delhi. The regional workshop was inaugurated by H. E. Kamal Nath, Minister of Environment and Forests, and the valedictory address was delivered by H. E. Salman Khursid, Minister of State for External Affairs. The national seminar was opened by Mr Tejinder Khanna, Secretary, Ministry of Commerce. Mr Erling Desau, Resident Representative of UNDP at New Delhi, also participated. The participants included experts from the South Asian Agreement on Regional Cooperation (SAARC) countries, representatives of the OECD and UNDP, officials from the various Ministries of Commerce and Environment, as well as several NGOs. Two experts from each of the SAARC countries were invited to present two papers each: one on the importance of free trade in promoting sustainable development, and the other on the mutual compatibility of trade and environment policies. The sectors covered by the national seminar included environmentally sensitive products such as textiles, leather, tea, refrigerators and other products which reflect the impact of environmental standards and concerns in OECD countries on the trading opportunities of India.

This book brings together the papers presented at the workshop and the national seminar. Almost all the papers have been authored by experts from the region, and it therefore constitutes an important contribution to capacity building as well as providing an insider's perspective on trade and environment issues for those outside the region. The workshop was organised under the aegis of a UNDP-funded project on Building Institutional Capacities for Multilateral Trade in the Asian Region. UNDP's support is gratefully acknowledged.

Thanks are also due to the many people whose assistance was vital to the publication of this book. These include the project coordinator, Mr Jagdish Saigal, the authors of the papers and the editors, Mr Grant Hewison and Ms Maree Underhill. We also acknowledge the contribution of Ms A. Achanta, Ms Veena Jha, Ms Susana Navarro, Mr P. Ghosh, Mr Roland Mollerus, Mr R. Pachauri, and Mr René Vossenaar in organising the seminar, writing the terms of reference for the papers, commenting on drafts and in handling the administrative tasks involved in putting this book together. Last but not least, mention must be made of Mr Vijay

Kelkar, former director of the International Trade Division at UNCTAD, who worked hard to ensure the success of the meetings and the publication of this book.

RUBENS RICUPERO
Secretary-General
United Nations Conference on Trade and Development

Notes on the Contributors

Amrita M. Achanta is a research associate at the Tata Energy Institute, Delhi. She is a recipient of the Mitchell International Prize for Sustainable Development, 1991, presented by the Houston Advanced Research Centre, Centre for Growth Studies, Texas, USA. Her research includes science and policy issues relating to the global environment, and in particular technology policy and intellectual property rights issues as well as biodiversity policy. She has a master's degree in zoological sciences from Delhi University.

Vasantha Bharucha is the Economic Adviser to the Ministry of Commerce, Government of India. Having obtained her PhD from a distinguished Indian university, she has worked for various international organisations and with the national government.

Achyut Bhandari is Director of the Policy and Planning Division of the Ministry of Trade and Industry of Bhutan, a position he has held since 1992. Prior to this he served as Director of the SAARC Division in the Ministry of Foreign Affairs and held several other positions in the same ministry between 1976 and 1986, including diplomatic posts at the Bhutanese Permanent Missions to the United Nations in New York and Geneva. He holds a degree in economics from the University of Western Australia.

Pradeep Dadhich is with the Tata Energy Research Institute, New Delhi, working in the area of industrial technologies relating to energy efficiency and cleaner production processes. His main interests include cogeneration and process simulation. He has a Bachelor of Technology in chemical engineering, and AIMA diploma in management and many years of industrial and research experience.

Ebba Dohlman has worked primarily on trade and trade-related issues since she joined the OECD in 1985. Having started in the Trade Directorate, she is currently in the Economics and Environment Division of the Development Co-operation Directorate. Prior to joining the OECD she worked as a consultant for GATT and also briefly for UNCTAD. She is Swedish by nationality.

Prodipto Ghosh is a Senior Fellow at the Tata Energy Research Institute, New Delhi. His professional affiliations include the American Economic Association and the Association of Environmental and Resource Economists, Washington, DC. He has been a member of the Indian Administrative Services since 1969. His research interests include the economic and policy aspects of global environmental issues, macroeconomic modelling, energy policy and technology policy. He has a PhD in economics from the Carnegie-Mellon University and a Bachelor of Technology in chemical engineering from the Indian Institute of Technology, New Delhi.

Lalith Heengama is a senior member of the Sri Lankan Administrative Service and has held the positions of Director of Plan Implementation, Director of Employment and Manpower Planning and Director-General of Sri Lanka Customs, and has been a member of the Presidential Commission on Tariffs. At present he is also an Additional Secretary and State Secretary to the Ministry of Trade and Commerce. He earned his Bachelor of Arts in Ceylon and is also a Bachelor of Law.

Grant Hewison is a Senior Lecturer on international relations, law and research at the Auckland Institute of Technology, New Zealand. His research includes the fields of trade and the environment, international law of the sea and international fisheries. He has written and edited a number of publications, including the edited publication *Freedom for the Seas in the 21st Century: Toward Ocean Governance and Environmental Harmony* and a publication titled *Reconciling Trade and the Environment: Issues for New Zealand*. He gained his MA in international environmental law at Auckland University, New Zealand.

Ashok Jha is with the trade section of India's Ministry of Commerce. He has been a civil servant with the Indian Government for twenty-five years, working in different capacities in various fields including finance and trade. He has also served as a Finance and Trade Counsellor at the Indian Embassy in Washington. He studied economics at Delhi University and at the Australian National University in Canberra.

Veena Jha is an Economic Affairs Officer at UNCTAD. She has served as a consultant for the United Nations Conference for Environment and Development (UNCED), the International Labour Office (ILO) and the United Nations Institute for Social Development. She has been a professor and researcher at Queen's College, Oxford, the University of London, where she completed her doctorate in economics, the Lady Spencer Churchill School of Management at Wheatley and the School of Economics at the University of Delhi.

Ramesh Jhamtani works with the Indian Planning Commission in the Environment and Forest Unit. His primary responsibility is to reconcile the environment and development in formulating medium- and long-term policies, plans and programmes. Earlier he spent over fifteen years dealing with different stages of industrial project analysis, including analysis within the social benefit cost framework. He has trained in project analysis at the World Bank's Economic Development Institute and he gained a Bachelor of Technology in civil engineering at the Indian Institute of Technology, New Delhi.

V. Kanesalingam of Sri Lanka was until recently Executive Governor of the Marga Institute, Colombo, and Secretary-General of the Committee on Studies for Cooperation in Development in South Asia (CSCD). He retired as Director General of Economic Affairs in 1972. He was thereafter a UN expert in development administration and later Professor of Public Administration at the Chinese University, Hong Kong. He is author of two books, *A Hundred Years of Local Government in Ceylon* and *Pricing Policy of Public Enterprises in Ceylon*, and has edited six books in a series of publications on regional cooperation in South Asia. He served as a member of the Sri Lankan government delegation to several sessions of the United Nations General Assembly, ESCAP and UNCTAD. He earned his PhD in economics in London.

Akhtar Hasan Khan of Pakistan serves as Chairman of the National Tariff Commission. During his thirty-odd years of public service he has worked variously in the Economic Affairs Division, the Ministries of Commerce, Production, Finance and Planning, and Development. He is the author of many articles on issues related to international trade, development economics, census and demography issues, public sector enterprises and social sector issues. He has a master's degree in economics from Karachi University, a master's degree in public administration from Harvard University and a PhD in economics from Tufts University.

Fasih Uddin Mahtab is a member of the Expert Group of Climate Change and Sea Level Rise and of the Expert Group on Environmental Concerns and the Commonwealth, both constituted by the Commonwealth Secretariat, London. In Bangladesh he has been a Cabinet Minister variously in charge of finance, agriculture and forestry, and planning. He has served on a number of national and international panels investigating various engineering and environmental issues and has published eight books on engineering and development problems. He has a PhD from Manchester University and is Rashid Professor at the Institute of Appropriate Technology, Bangladesh University of Engineering and Technology, Dhaka.

Roland Mollerus studied economics and international relations in the Netherlands and Canada and obtained an MA in international political economy. At present he works in the Trade and Environment Section of UNCTAD. For several years he has worked with a trading company in the Netherlands. He has also served as a consultant for the Permanent Secretariat of the Latin American Economic System (SELA), the Ecological Management Foundation (EMF) and Jaycees International.

Kamal Nath holds the post of Minister for the Environment and Forests in the Indian government, a post he has held since 1991. He was first elected to parliament in 1980. Since taking charge of the Ministry of Environment and Forests, Mr Nath has overseen the development of the National Policy on Ecological Conservation and the Abatement of Pollution, the establishment of environmental tribunals, the introduction of environmental auditing and an Indian eco-mark or green labelling scheme. He has also encouraged afforestation and development of degraded wastelands in India. Kamal Nath has led Indian delegations at a number of international conferences on the environment, and emerged as a chief spokesperson for developing countries at the 1992 United Nations Conference on Environment and Development. He holds a bachelor's degree in commerce from St Xavier's college, Calcutta.

Ligia Noronha of India is a fellow at the Tata Energy Research Institute. Her research includes development economics, the interface between environment and development, natural resources policy and use, the political economy of North–South issues, technology transfer, international environmental policy and regime formation. She has a PhD from the London School of Economics.

Khilendra N. Rana is Executive Director of the Institute for Sustainable Development in Nepal. He has worked in various capacities in different countries and was a member of the Royal Nepal Academy of Science and Technology. He has also served as the Nepalese national liaison for science and technology issues in several national and international organisations. He has a master's degree in mechanical engineering and a masters degree in economics.

Rakesh Shahani has research experience at the Indian Council for Research on International Economic Relations at HARIDON and at the Institute of Economic Growth. He has a bachelor's degree in commerce and a postgraduate degree in business economics from Delhi University.

M. Nevin Shaw is a Canadian national who was attached to the International Institute of Sustainable Development at Winnipeg, Canada. He has occupied various positions of responsibility with the Canadian government.

Preeti Soni is a research associate at the Tata Energy Research Institute, New Delhi. Her research is in the area of international trade, environmental issues, and economic and energy policy. She gained her master's degree in economics at the Delhi School of Economics.

Ana Paola Teixeira is a Brazilian economist, currently working at UNCTAD. She has worked as a consultant for the UNDP and the University of Campinas, Brazil, and she lectures at Webster University in Geneva. She graduated from the American University of Paris and took a master of philosophy degree in development studies at the University of Sussex.

Maree Underhill teaches English literature and language in Auckland, New Zealand. She has edited and sub-edited a number of publications. She gained her bachelor's degree in political science and English literature at Auckland University, New Zealand.

M. C. Verma works in the areas of trade and the environment and development and trade, among others for the Indian Administrative Service. Earlier he held senior positions within the Indian Administrative Service, including Human Resource Development Planning. He has been a consultant to the United Nations in various African countries and was Chief Technical Advisor of a UNDP-ILO project in Nigeria. He has prepared numerous papers for international organisations, including the World Bank and the UNDP. He gained a master's degree in statistics, with econometrics, at the University of Luknow, India.

René Vossenaar is an economist from the Netherlands. He is Chief of the Trade and Environment Section of the International Trade Division of UNCTAD. Previously he was a researcher at the University of Tilburg in the Netherlands and worked for several years with the Economic Commission for Latin America and the Caribbean (ECLAC) in Santiago, Buenos Aires and Brasilia. He has been with UNCTAD since 1985.

Christine Wyatt is a sociologist and economist who has worked with UNCTAD on trade, environment and technology issues. Her work on international trade also includes technical cooperation in export promotion and trade liberalisation, and she has written a number of articles and a book on regional integration. Currently Christine Wyatt is conducting research on the socioeconomic impact of environmental measures in developing countries.

Introduction: Trade, Environment and Sustainable Development: A South Asian Perspective

Grant Hewison and Maree Underhill

The linkages between the objectives of liberalised international trade, protection of the environment and sustainable development have become extremely important to international policy making. Although the conclusion of the Uruguay Round of the GATT negotiations appears to have ushered in a new era of more open and expansive international trade, there are concerns about a collision between freer trade and the policies being utilised to protect the environment. While the concept of sustainable development incorporates a requirement to make trade and environment objectives mutually supportive, like the concept of sustainable development itself, this raises more questions than it immediately answers. Many developing countries have raised concerns that the promises now being made to integrate trade and environment policies will be illusory and undermine the achievements of the Uruguay Round and their recent efforts to liberalise their economies.

This book seeks to clarify many of these concerns and identify some of the paths by which all countries can move forward. It does so from a unique perspective – that of developing countries, and in particular the developing countries of South Asia. Almost all the authors who have contributed chapters to this book are from countries in the South Asian region, and all are experts in the field of trade and the environment. The book also builds on the work being undertaken by the United Nations Conference on Trade and Development in their efforts to assist developing countries negotiate an international environmental policy framework that does not have an adverse impact on trade.

TRADE AND THE ENVIRONMENT: A SOUTH ASIAN PERSPECTIVE

Many economists contend that if all the current external environmental costs were internalised and sound environmental policies instituted to enforce this process, then trade and environment policies would not conflict with each other. But this appears, at least in the short term, to be only a dream. Full cost internalisation is rarely possible and sound environmental policies have not yet been universally implemented. Consequently we are left in a situation where some countries internalise their environmental costs more fully than others, with international trade and investment tending to exaggerate these differences.

Environmentally orientated regulations, standards and voluntary mechanisms have created barriers to trade and look set to continue to do so in the future. Although these mechanisms are generally undertaken for genuine environmental purposes, opportunities also exist for them to be used in a protectionist manner. Regulations, standards and voluntary mechanisms are also being used to shift the environmental behaviour of firms, not only with regard to the end products they produce, but also the process and production methods (PPMs) used during the manufacture of products. Indeed, from an environmental perspective, it is often these PPMs that damage the environment rather than the final products themselves.

At present countries are not permitted, under international trade rules, to treat products differently simply because their PPMs produce different environmental outcomes. Thus tea grown with an excessive use of pesticides or fertilisers cannot be treated differently by governments from tea grown organically. Nevertheless private or voluntary initiatives, such as eco-labelling and eco-packaging, that target the 'green consumer' and are tolerated by current international trade rules are beginning significantly to alter the market access and competitiveness of 'green' products over conventionally manufactured products, particularly in specialised or environmentally conscious markets. These types of voluntary market-orientated mechanisms are also starting detrimentally to affect the market access and competitiveness of imported products compared with products manufactured locally. The costs of joining eco-labelling and eco-packaging schemes, altering production or packaging to comply with the schemes or arranging for certification can be considerable, especially if the criteria used differ from one export market to another. For firms in developing countries, these issues have to date appeared only to create small disruptions, but the future does not seem so promising. These eco-labelling, eco-packaging and other environmental schemes are now directing their attention towards products such as textiles, clothing and footwear that are manufactured predominantly in developing countries and will without doubt have serious effects on their future international trade.

Although the environmental costs of production are not equal in all countries, because of differences in assimilative capacities as well as variations in social and cultural preferences, the regulations, standards and voluntary measures being adopted tend to establish criteria based on the goals of the most developed countries. This impacts even further on firms in developing countries who, in order to maintain their market access and competitiveness, are required to improve their environmental performance so that they meet these higher standards. On the other hand, as many of the authors in this book recognise, liberal trade policies without corresponding environmental protection can lead to serious environmental degradation, undermining future productivity and human welfare. Good environmental policy and performance by firms in developing countries may also provide opportunities to tap into the environmentally orientated consumer markets of the North.

The need for technical cooperation and assistance in understanding how environmental policies operate and their potential trade impacts will be very important for developing countries. In addition the ways in which environmental standards can provide developing countries with trading opportunities also need to be further explored. The use of complementary measures such as sound environmental policies and access to environmentally sound technologies may also provide keys to making trade and environment policies mutually compatible in developing countries.

PART I: ISSUES

The first part of this book explores a number of important overarching issues facing developing countries within the context of trade and the environment. While Kamal Nath's chapter examines these issues broadly and identifies a number of difficult political questions that will need to be answered, the chapters by René Vossenaar and Veena Jha, Christine Wyatt and Roland Mollerus focus more directly on how environmental initiatives in OECD countries affect the products and production processes of firms in developing countries.

In Chapter 1 Kamal Nath, India's minister for the environment, provides an overview of the many issues covered in more detail later in this book. Nath squarely places the trade and environment debate within the context of sustainable development and the outcomes of Rio. It is not enough, in Nath's view, simply to address the barriers environmental policies can place in the way of trade since the problems raised by trade and the environment are 'systemic of the entire international economy'. Nath contends that any resolution must also redress the imbalances in trade and debt between North and South. For developing countries it is poverty and underdevelopment that produce unsound environmental outcomes, and only through genuinely free and open trade will poverty and underdevelopment be overcome. Although freer international trade brought about through the Uruguay Round agreements may alleviate poverty in the South, Nath expresses the concern held by many other contributors to this book, that the environmental policies currently being promoted by the North are actually protectionism in 'green wrapping paper' and will undermine the achievements of the Uruguay Round. Nath ends with a note of caution for the North. Developing countries are not ready to allow the WTO or any other forum to review their national developmental priorities. Action taken to protect the environment must come about through free will in an atmosphere of shared global concern, not by the use of unilateral trade restrictions.

In Chapters 2 and 3 René Vossenaar and Veena Jha of UNCTAD examine the range of environmental regulations and standards being adopted by OECD countries and their effects on market access and competitiveness, especially for firms from developing countries. Vossenaar and Jha begin by making a distinction between policies that regulate process and production methods (PPMs) and policies that regulate the characteristics of a product itself. They argue that while PPM policies are essential for protection of the environment, there is little justification for applying these standards to imported products, since almost all environmental problems caused by PPMs are local and are not transmitted by the product. Moreover these authors contend that policies seeking to address PPMs tend to be based on a country's own particular social, cultural and development goals, which may be inappropriate for other countries. They make the same point regarding international harmonisation of environmental policies, and reject claims that variations between environmental policies in different countries affect competitiveness or result in the migration of 'dirty' industries. Although Vossenaar and Jha firmly reject the use of trade restrictions as a means of raising environmental standards, they are of the view that positive means, such as financial assistance, the transfer of technologies, the application of 'green' trade preferences, eco-labelling and sustainable production of commodities may appropriately be employed to raise standards. In their chapter dealing with environmental policies that regulate the characteristics of a product itself, they particularly focus on concerns that competitiveness and market access are being detrimentally affected by eco-packaging and eco-labelling policies. They argue that these policies impose more costs and raise more difficulties for imported products than products manufactured domestically and consequently create barriers to trade. Vossenaar and Jha point out that producers in developing countries will become increasingly exposed to these policies as OECD countries extend them to include products, such as footwear and clothing, that are now predominantly manufactured in developing countries.

Christine Wyatt's chapter (Chapter 4) presents a detailed case study of the effects of Germany's environmental policy making, eco-labelling and eco-packaging schemes on products imported from developing countries. Wyatt notes that although most attention has been directed towards the Blue Angel eco-labelling scheme, the more recent eco-labelling initiative for textiles appears to be even more troublesome for developing countries. Wyatt argues that textile manufacturers in Germany see this label as a means of capturing market share and offsetting the low labour costs enjoyed by firms in developing countries. Probably of equal concern is the German Packaging Ordinance, which obliges producers to take back all product packaging. The efficacy of this ordinance as well as the Dual System and Green Dot schemes established to comply with it are closely scrutinised. Of particular note for exporters from developing countries is Wyatt's discussion of the impacts of recycling transport packaging waste. She notes concerns about the huge exports of collected recyclable waste and other recycled products from Germany and argues that Germany is the world's leading exporter of waste. Although these measures may pose significant barriers to trade, Wyatt also notes that significant opportunities exist for exporters from developing countries who can demonstrate that their products are 'clean and green'.

Roland Mollerus' chapter (Chapter 5) examines the linkages between environmental standards and international competitiveness. He notes that it is common to believe that the need to comply with stringent domestic environmental process standards will adversely affect domestic industries vis-à-vis foreign competitors, if the standards are lower in the foreign country. It is feared that this is leading to the relocation of industries from countries with high environmental process standards to countries with lower ones or that a competitive advantage accrues to industries that operate in countries with lower environmentally orientated process standards. Mollerus first quantifies the significance of these so called 'polluting industries' for the exports of SAARC countries, and then goes on to examine whether indicators of international competitiveness demonstrate increased or decreased competitiveness for 'polluting industries'.

PART II: COUNTRY CASE STUDIES

In the second part of the book, a number of authors from the developing countries of South Asia offer their views of the effects of environmental policies on freedom to trade and the effect of liberalisation of trade on the environment. The South Asian region comprises Bangladesh, Bhutan, India, the Maldives, Nepal, Pakistan and Sri Lanka. All these countries share historic, geographical, cultural and economic links as well as similar environmental problems and trade patterns. While all the authors are concerned that the environmental policies being implemented by OECD countries can have negative effects on their countries' trade, and that at times these policies seem to have protectionist intent, they also recognise potential opportunities for the export of 'green' products to OECD countries. What is required is more information about these environmental policies and assistance in realising 'green' trade opportunities. From their own countries' perspective, most authors agree that the fundamental question is how best to alleviate poverty and at the same time reduce stress on the environment. While more open trade has been recognised as an important mechanism for achieving both these objectives, the authors generally contend that environment and poverty alleviation policies must be fully incorporated into trade policies and not just tacked on the end or left to wait for the wealth to trickle down.

In Chapter 6 Prodipto Ghosh, Preeti Soni, M. C. Verma and Rakesh Shahani describe the major environmental concerns facing this region and some of the mechanisms being employed to overcome those concerns. To develop economically, these authors argue that South Asian countries will need to cooperate with each other and with countries outside the region across a range of policy-making areas, but especially in the area of trade. Indeed recently there have been signs of cooperation. A South Asian Preferential Trading Agreement has been negotiated under the umbrella of SAARC. Both these agreements provide a number of avenues for further cooperation and more extensive interregional trade. Perhaps the lessons of Southeast Asia, where interregional trade has promoted significant growth, could be learnt by other regions. Although these authors are generally supportive of increased trade, they are worried that policy makers remain largely ignorant of the environmental degradation that can be caused by unsustainable development activity. A rapidly increasing population and recent industrialisation have, in the view of these authors, caused degradation of arable land, pollution of inland waterways, deforestation, alteration of coastal and marine environments and increased pressure on urban environments. Unless plans designed to alleviate these environmental problems are incorporated into trade and economic policies, the authors are concerned that development in the region will continue to degrade the environment.

In Chapter 7 Dr Fasih Uddin Mahtab of Bangladesh provides us with the first of the individual country case studies. He describes a number of trade and environment issues that arise out of leather production in Bangladesh and notes that recent trends in worldwide leather production point to a transfer of production from developed to developing countries, where low labour costs have become a comparative advantage. Although Bangladesh has, in his view, made modest efforts to establish itself as an important producer of leather, several technological and environmental concerns remain to be resolved. Dr Mahtab argues that while environmental management systems appropriate to Bangladesh's overall conditions should be developed, these efforts must take into account the question of international competitiveness so that they do not undermine the potential of the leather industry to contribute to the Bangladesh economy.

Achyut Bhandari of Bhutan offers the second country case study (Chapter 8). Bhandari notes that while Bhutanese trade will continue to rely on the export of electricity, timber, wood products, minerals and some agricultural products, there are opportunities for niche markets in products such as traditional handicrafts, mushrooms, essential oils, flavours and herbs. He argues that Bhutan's relatively unpolluted environment may prove to be a positive factor in locating markets for some of these products. However he notes that trade contacts for these 'greener' products can only be made with bilateral and multilateral assistance. Bhandari completes his study by pointing out that Bhutanese trade, at least in the near future, is likely to concentrate on the South Asian region and that any problems associated with trade and the environment will have to be tackled first at the bilateral or regional level, while keeping in mind developments at the global level.

Three authors provide case studies from India. R. C. Jhamtani provides us with the first Indian case study (Chapter 9). He notes that there are a number of factors, set at different political, economic and environmental levels, that will influence the potential for conflict between India's trade objectives and the environmental impact of those objectives. These include political processes within the WTO, economic factors such as India's balance of payments position and trade liberalisation, as well as the impact of industry on India's own internal environment, particularly where goods are being produced for export. Jhamtani argues that the acute scarcity of financial resources and the reluctance by OECD countries to transfer environmentally sound technologies are important constraints on directing developing countries along the path to sustainable development. But reconciliation of trade and environmental issues, argues Jhamtani, will not just require the transfer of technology and assistance. It will require a new set of international goals rooted in global ethics and international equity.

Ashok Jha offers the second Indian case study (Chapter 10) and focuses on India's leather exports. Jha argues that environmental concerns and trade are not *ipso facto* contradictory. Since trade is a recognised facilitator of economic development, growth in trade should have a favourable impact by increasing income, which can then be expended on the environment. Jha is concerned that environmental protection will be used by the North as a non-tariff barrier to trade and reduce income opportunities for the South that might have been spent on environmental improvements. He contends that an open, non-discriminatory trading environment backed by WTO rules is a better system for protecting the environment than policies that impinge on free trade.

Dr Vasantha Bharucha offers the final case study of India (Chapter 11). Dr Bharucha examines the impact of a number of international and country-specific environmental measures on a selected range of India's exports. Included in the survey are the refrigeration and air conditioning industry and its use of ozone-depleting substances, as well as manufacturers of tea, dyes and intermediates, agricultural products and processed foods, marine products, and the leather and textile sector. Dr Bharucha concludes that there is no doubt that Indian exports will confront new problems relating to the environmental standards set in foreign markets as well as standards set by multilateral agreement, such as those in the Montreal Protocol on substances that deplete the ozone layer. In Dr Bharucha's view, however, it will make more economic sense for India to incur these trade-related environmental costs in the short term to ensure continued access to rich overseas markets over the long term. As India, along with many other developing countries, has liberalised its economy and pursued trade-related growth, in Dr Bharucha's view there can be no choice but to conform to the new environmental standards being set. There will, however, be a need for international funding, overseas aid, transfer of technology and foreign investment to assist Indian industries during their transition to environmentally clean production processes.

In Chapter 12 Khilendra N. Rana provides us with a case study from Nepal. Like Bhandari of Bhutan, Rana observes that the emerging consumer preference for 'green products' in OECD countries offers important opportunities as well as possible difficulties for Nepalese exports. Although the need to feed an ever-increasing domestic population will require continued application of chemical fertilisers, pesticides and growth stimulators, argues Rana, the emerging consumer preferences in OECD countries clearly call for a switch to 'green products', at least in the export sector.

In Chapter 13 Dr Akhtar Hasan Khan of Pakistan identifies three central issues to making trade and environmental policies mutually compatible. He identifies environmentally orientated import restrictions in countries importing Pakistani products as the first issue. The second issue relates to the aggressive pursuit of exports in Pakistan and whether this is leading to environmental degradation, while the third he considers is whether trade or tariff policies can be used to foster environmental ends. With regard to the first, Dr Khan is concerned that environmental issues are becoming disguised restrictions on international trade and are providing ammunition to protectionist lobbies in all countries. He argues, with regard to the second issue, that OECD countries have already passed through an earlier polluting stage of industrialisation and cannot now legitimately choke the development of late comers by insisting on standards that they themselves did not conform to when they were at a similar stage of development. Finally, trade and tariff policies can, he believes, be used as a tool for promoting environmental ends where clear environmental targets are selected and trade measures are used in conjunction with other measures.

Lalith Heengama and Dr V. Kanesalingam provide us with two separate but similar case studies from Sri Lanka (Chapters 14 and 15). Both authors note that sustainable development will not be achieved if its very basis is undermined by environmental degradation caused through liberalised trade. They argue that degrading soils, depleting resources and destroying ecosystems in order to raise incomes in the short term will inevitably undermine future productivity and income. In their view there is no question that trade and environment policies must be made mutually compatible, not least because a merging of both will be required to increase welfare. To bring about mutual compatibility between these policy objectives, they argue that there must be a complete understanding between all countries engaged in international trade and the internalisation of current external environmental costs.

PART III: PATHS FORWARD

The third part of this book focuses on some of the paths forward from the present conflict between trade and environment policies to a future policy that is mutually supportive of both. In particular, this part again attends to the special concerns of the developing countries of South Asia and focuses on the transfer of environmentally sound technologies and the provision of aid.

Veena Jha and Ana Paola Teixeira of UNCTAD note in Chapter 16 that the internalisation of current external environmental costs will require dramatic changes to the technology presently used in developing countries. Part of the solution will lie in the development and transfer of more environmentally sound technologies (ESTs) from the North to the South. They observe, however, that there are several barriers to prevent developing countries from gaining access to ESTs. In the first place, there are concerns that the reasons for past reluctance to transfer ordinary technologies will mean the same fate for ESTs. In addition, the selection and application of ESTs for developing countries will have to be handled very carefully if they are to be implemented successfully. There are also concerns about the viability of ESTs if they are being applied only to the export sector. Finally, Jha and Teixeira note that costs, patenting and licensing controls, and to a lesser extent the way standards are formulated in some OECD countries may also act as barriers to EST transfer. The view of these authors, however, is that the transfer of ESTs from developed to developing countries will be essential to sustainable development, and that a mechanism must be found to make their transfer simple and successful.

In Chapter 17 Amrita N. Achanta, Pradeep Dadhich, Prodipto Ghosh and Ligia Noronha from India offer another perspective on the transfer of environmentally sound technologies. They have directed their attention to the many lessons that can be learnt from the difficulties involved in the transfer of conventional technologies that might assist in developing better policies for the transfer of ESTs. They point out that policy making must address issues at the levels of the economy, the sector and the individual firm. Existing barriers to the diffusion of technology, such as government taxes and administrative red tape, need to be removed. Firms in developing countries must, they argue, be encouraged to enter joint ventures and appropriate technology agreements as well as undertake appropriate training, if the transfer of ESTs is to be successful. The greatest failure of technology transfer policy in the past has been the inadequate attention paid to indigenous knowledge, indigenous technology and the creation thereof. While the authors conclude that the transfer of ESTs will be essential to sustainable development, they are concerned that unless improvements are made to existing technology policy, the same difficulties faced by conventional technology transfer will undermine the transfer of ESTs.

In Chapter 18 Ebba Dohlman of the OECD focuses on the role that aid agencies can play in promoting compatibility between trade and environment policies. She notes that aid agencies can assist developing countries to take up new trading opportunities arising out of environmental policies or consumer preferences, as well as overcome trade constraints resulting from increasingly stringent and complex environmental regulations. Dohlman argues that the provision of timely and comprehensive information about environmental opportunities, regulations, and possibilities for financial and technical assistance is very important and a natural function of aid agencies. While the catchphrase of developing countries in the past few years has been 'trade not aid', Dohlman contends that there is still an important place for aid agencies and aid programmes in assisting developing countries to meet their potential as environmentally kind trading nations.

Nevin Shaw of the International Institute for Sustainable Development (Chapter 19) argues that we must place the trade and environment issue within the broader context of finding new ways of integrating domestic and international economies and environments into our decision making at all levels. This has to be done by taking account of our common but differentiated responsibilities for past and future uses of planetary resources and with an understanding of the issues of equity, poverty alleviation and lack of capacity among many of the members of the international community. To assist the global community in achieving these objectives, the International Institute for Sustainable Development has developed a set of trade and sustainable development principles that are outlined in this chapter.

The concerns held by developing countries that the commitments made during the Uruguay Round to liberalise international trade will prove illusory are very serious. There is widespread belief among many developing countries that new trade barriers, masquerading as environmental protection policies, are being erected in developed countries. While these same developing countries recognise that protection of the environment is a priority, they are not ready to allow the WTO or any other forum to reassess their national environment and development plans. Action taken to protect the environment, they argue, must come about through free will in an atmosphere of shared global concern, not through the use of trade restrictions imposed unilaterally by countries with the largest markets. Moreover, from the perspective of these developing countries, the trade and environment debate goes much further than simply trying to overcome the trade distortions brought about by environmental policies. It is about achieving sustainable global development and equity between all peoples.

Any resolution must redress the imbalances in trade and debt between North and South.

In the final analysis, resolution of trade and environment issues will require education, confidence building, negotiation and compromise on the part of all involved. It is hoped that this book will assist in this process.

Part I

Issues

1 Trade, Environment and Sustainable Development

Kamal Nath

The concern labelled 'Trade and Environment' is not so much about the linkages between liberalised international trade and environmental degradation, as about the whole question of sustainable development and how this type of development can be furthered by international commerce. Indeed we would be starting off on the wrong foot if we were to ignore the advances made at Rio in translating the poetry of environmental protection into the practicality of sustainable development. The realisation that international trade and sustainable development are not incompatible is a significant advance from the earlier debate, which presumed that international trade and development could only be at the expense of the environment.

Sound environmental policies can help secure the gains from international trade, while at the same time greater trading opportunities can help developing countries to invest more in environmental protection. Restrictive trade measures do not appear to be the best way of addressing environmental concerns. The roots of the linkage between trade and the environment are not to be found in superficial assumptions since they go much deeper and are systemic of the entire international economy. Indeed it is a retrograde way of shaping environmental norms to allow dispute resolution panels of the WTO acting on a case-by-case basis to steer global trade and environmental policy making.

Furthermore it is important to recognise the social and environmental subsidy that industrialised nations receive from developing countries. This insidious subsidy renders all development in the North unsustainable by definition. It makes a mockery of free trade. If we are to set things right, then the subsidy must be recognised and accounted for. The economic poverty and environmental degradation that afflicts the South is not simply coincident with the affluence of the North. It is poverty and degradation visited upon the South *by* the affluence of the North. Banishment of this impoverishment is crucial to everyone's survival. To achieve this, the debate about trade and the environment must go beyond superficialities.

Unfortunately, however, the debate seems to be stymied by the subtle authoritarianism of economics.

Although Agenda 21 is a vast and wordy document, the tendency to gloss over it must be resisted, since it presents a consensus between nations about the crucial issue of sustainable development – a consensus that has been achieved on a scale unparalleled in human history. Agenda 21 explicitly recognises that 'an open, equitable, secure, non-discriminatory and predictable multilateral trading system is a benefit to all trading partners'. Consequently any decisions made to modify trade relations by incorporating environmental considerations must conform rigorously to this consensus. It is obvious that the environment is something the whole world is concerned about, and that disaster would affect us all, whether rich or poor. It is also readily apparent, however, that present trade relations are so skewed and the imbalances so glaring, that without structural correction of these imbalances any linkage between trade and the environment is very likely to work to the detriment of the poorer countries.

India as well as several other developing countries and countries whose economies are in transition, have each in their own way made considerable strides towards liberalising their economies. However the economic and environmental benefits of trade liberalisation will only be cemented if market forces are permitted to work and the true costs of products are recognised in their price. At present this is not the case. In the absence of appropriate pricing, developing countries are being forced to overexploit their natural resources, with liberal trade policies only hastening the process. However the solution is not to revert to a tightly controlled trade regime, but instead to incorporate environmentally related costs into international market prices through refining and extending the well-known 'polluter pays principle' into a 'buyer pays principle'.

India is among the vanguard countries of environmental protection. India has established environmental standards for its own products and processes as well as environmental impact assessment procedures for the location of industries and the technology used by them. Environmental auditing procedures and an eco-labelling scheme have been instituted to India.

But how can a developing country that seeks voluntarily to move even further along the path of environmental protection expect to have a comparative advantage in trade if it continues to be unduly exposed to the risks of environmental underpricing? There are historical differences to contend with and a significant effort will be required to solve the urgent problems of poverty and economic underdevelopment. If trade advantages were also to recede, because of restrictions under the garb of environmental protection, this would not only slow down India's development efforts, but also aggravate its environmental problems. In poor societies, environmental degradation is the manifestation of poverty. It is poverty that is the enemy. The extent to which a poor society is restricted from reaping economic advantage through international trade, is the extent to which the condition of poverty is prolonged.

Much is heard about the dangers of protectionism, yet some developed countries are reluctant to abandon their own protectionist policies. Because protectionist-driven regulations cannot be applied in a obvious manner, they now appear instead to be packaged in 'green wrapping paper'. This can occur when developed countries specify standards for manufactured items, or worse, for the processes by which the items are manufactured, that firms in developing countries cannot meet. For instance some eco-labelling schemes in Western countries give value only to environmentally friendly chemical dyes and ignore natural dyes completely, even though they are equally, if not more, ecologically sound. Asian countries that are major manufacturers and exporters of textiles, but who often use natural dyes, may not be able to earn an eco-label for their products. In order to comply with standards, manufacturers in developing countries may have to purchase the technology of countries that are potential importers, while under this guise the importing countries are finding a market for their technology, which may not be the most appropriate for the developing country concerned. These types of protectionist restrictions can limit production in developing countries, prevent access to lucrative markets, or even compel the purchase of specific environmentally orientated technologies from developed countries.

There is also scope for importing countries to set standards that relate to the processes by which exported goods are produced, thus very subtly influencing the national policies of the exporting countries. However different countries have different environmental objectives and should be able to take their own path in order to achieve these objectives. Timber is particularly sensitive example. In India, forests are not worked or logged commercially, India's commercial timber requirements being met by imports. In spite of this, India cannot accept forest conservation schemes that would lock up India's forests. Primarily, this is because forests in India are a community resource with hundreds of millions of people depending upon them for fuel, fodder, medicine and fruits. These communities by and large use the forests sustainably, even if this type of use does not easily fit into the Western idea of conservation. This highlights the fact that there can be no absolute environmental standards or specifications, but that these are specific to all countries or communities.

India agrees that environmentally harmful production processes must be phased out and that overexploitation of non-renewable resources must be controlled. However the solution lies not in unilaterally banning trade, but rather in transferring technology and paying appropriate prices that reduce the need for overexploitation. Moreover, systemic problems cannot be solved in isolation. While child labour in the carpet industry may draw attention because carpets are exported, it may be better for a poor country to direct its limited resources into rehabilitating children working in more hazardous industries. Nevertheless, in order to overcome trade barriers the country may be forced to rearrange its development priorities to its own economic and social detriment.

In fact the whole concept of international eco-labelling based on the processes by which products are manufactured amounts to a legitimisation of extraterritorial interference by one country over another's domestic affairs. It is a kind of green imperialism.

Environmental effects may be reversible or irreversible. They may be confined to national boundaries or spill over them. Reversible effects should not unduly disturb the international community because they are actually a reflection of the considered national priorities that a country sets for itself on its path towards economic and social development. Similarly, environmental effects that do not cross national boundaries should be the sole concern of domestic authorities. They should be left to deal with them in accordance with their own policies and be commensurate with their own technological and financial capabilities. Anything beyond this would not only be an infringement of national sovereignty, but would certainly smell of rank protectionism. However, when environmental effects do cross over national boundaries, those nations that are affected certainly have the right and duty to safeguard their own vital concerns. But if left unchecked, unilateral actions to protect the environment could bring the international trading system to a chaotic halt. The temptation to apply domestic legislation extraterritorially must be guarded against. Cases of transboundary pollution or other environmental effects should be dealt with through genuine multilateral mechanisms that fully represent all geographical as well as developmental interests. Cautious and broad-based consensus must be the order of the day.

Even product standards can become trade barriers. It would be unrealistic, and perhaps undesirable, to say that there should be no standards whatsoever. It is also theoretically desirable that product standards should be harmonised, since if each country decides to have its own standards, the resulting confusion would be disastrous. In a free-for-all, only the rich and powerful would succeed! While harmonisation of product standards is desirable, there seems little possibility of setting single standards at the global level because different countries can sustain different ambient levels of pollution, have different environmental assimilative capacities and different social and economic objectives. However some harmonisation of standards may be possible, either by product category or on a regional basis, or perhaps better still, through an innovative combination of both of these approaches.

But without a concomitant transfer of technology and finance there seems little chance of developing countries changing their production so as to reach these standards. Technology is the key – clean and state-of-the-art. This takes us back to Rio and UNCED. Unless there is concerted effort on the part of developed countries to transfer clean technology to developing economies, what is the use of discussing standards at all, let alone harmonising them? In spite of low levels of capital and antiquated technology, developing countries are doing their best to live sustainably. If every person on the planet used fossil fuels at the rate of the average citizen of the North, we would all have boiled long ago! Conversely, if every person on the planet used fossil fuels at the rate of the average citizen of India, climate change would not be an issue at all.

CONCLUSIONS

The South as a whole *favours trade* and is not seeking to *trade favours*. A moratorium should be placed on linking trade with the environment, unless an environmental issue clearly affects collective ecological security. In that case it should be dealt with multilaterally, not unilaterally. Protectionism should not be passed off as environmental concern. Developing countries are not ready to allow the WTO or any other forum – trade or otherwise – to review national developmental priorities. The point was made at Rio and has repeatedly been made at the Commission on Sustainable Development. Action taken to protect the environment must come through free will in an atmosphere of shared global concern. The results of these efforts would be even better if trade was expanded in a way that addressed the amelioration of poverty.

Historically the hand on the South has been harsh. The intergenerational equities spoken of in current environmental debates should also consider past intergenerational inequities. While present generations must not appropriate from future generations, the misappropriations by past generations cannot be brushed aside. The discussions that will take place concerning trade and sustainable development will be very difficult and will occupy the future agendas of many fora. But whatever words are used to sugar-coat or explain it away, the fact remains that trade has had a sharply iniquitous edge. This edge must not be allowed to sharpen through the taking of hasty decisions. The nexus between trade and environmental policy must be mutually supportive and not work to the detriment of them both.

2 Environmentally Based Process and Production Method Standards: Some Implications for Developing Countries

René Vossenaar and Veena Jha*

Environmental standards, in the context of international trade issues, raise important questions. How do environmental policies, standards and regulations affect market access and competitiveness? How can countries pursue environmental policies appropriate to their circumstances within an international trading system? What relationship should there be between the trade measures used in multilateral environmental agreements and the rules of the international trading system?

The impact of environmental standards on market access and the possibility of these becoming non-tariff barriers to trade is a special concern for developing countries. Traditionally attention has focused on the *product standards* issue. Exporters from these countries are anxious that their products will be denied access or that they will incur high adjustment costs in order to maintain access to overseas markets.

Although attention in the past was largely directed at the potential negative trade effects of product standards, from an environmental perspective both product standards and standards relating to process and production methods (PPMs) need to be addressed. This is because environmental problems are more often associated with the production process than with the product itself. Moreover these PPM-related standards are becoming more stringent and comprehensive in developed countries, in response to an improved understanding of environmental risk and public preference for tighter environmental protection.

While the use of PPM-based standards for environmental purposes is essential, PPMs need not be a serious trade issue. Since most environmental problems are intrinsically local, their effects not being transmitted through the product to the environment of the importing country, there is little justification for applying PPMs to imported products. Furthermore, current international trade rules support the right of countries to set PPM standards that are appropriate to their own environmental and development priorities.

Where PPMs are expressed in the characteristics of the final product, however, they are, in principle, subject to the same international trade rules that are applied to any other product standard. In cases where a multilateral environmental agreement applies to both product and PPM-related trade measures to achieve a multilaterally agreed environmental outcome, the distinction between these measures becomes less of a problem, except perhaps where the environmental concern relates to an ethical, value or cultural preference intrinsic to the PPM.

Although rigorous process standards generally have positive effects on sustainable development by requiring the internalisation of external environmental costs and by removing inefficiency, concerns about maintaining competitiveness may turn a country away from applying rigorous standards. The enforcement by one country of its own PPM standards on products imported from another country, however, is an inappropriate and inequitable way of achieving sustainable development, particularly if the latter country is an economically weaker partner.

This chapter analyses the impact of PPM-based standards and regulations on market access and competitiveness, particularly from the perspective of developing countries. First, it is worthwhile defining PPMs, the distinctions between process- and product-based regulations and their effect on trade. The chapter then focuses on those PPM standards which address impacts on the local environment and are not transmitted by the traded product. The principal concerns here are friction over 'hidden' environmental subsidies and 'eco-dumping', which could in particular target competitive developing countries; pressures for harmonisation of PPM-based standards, which may be inappropriate for developing countries; PPM criteria based solely on the priorities of OECD countries; and the costs incurred by developing country exporters in upgrading their PPMs and providing information about the environmental impact of their products.

PRODUCT AND PPM STANDARDS

A distinction must be made between process and production method standards and product regulations.[1] Whereas product standards lay down the criteria a product should meet, such as performance, quality, safety or dimension standards, process and production method standards specify how the products should actually be produced. Product regulations, on the other hand, try to control the effects that products have in the place where they are consumed. If a product is traded internationally, these effects will take place in the domestic environment of the importing country. Under the WTO rules, a country may require that imported products comply with the same product regulations as domestically produced products, with a rider that these regulations can be challenged if they constitute unnecessary obstacles to trade.

Although different categories of PPM standard exist depending upon the location in which the environmental effect they seek to address takes place, this chapter will focus mainly on PPM standards that address environmental effects in the producing country.[2] This is because pressures for PPM-related trade restrictions are, in these cases, based on competitiveness concerns and the need to comply with domestic environmental standards rather than on the environmental impact made during consumption. The risk of protectionism is particularly high in these cases and the pressures for trade restrictions are also highly controversial. Moreover, although PPM-based regulations cannot be formally applied to imported products because the current WTO rules do not permit this, the difficulties of complying with non-mandatory PPM-related standards and consumer preferences in these markets are having an impact on export competitiveness.[3]

PPMs related to the product

When PPMs are related to the product, the use of PPM standards may have practical advantages over product standards. In certain cases it may be more efficient and more economical to control the process rather than the product itself to ensure product quality. It has been observed that:

> Traditionally manufacturing standards have emphasised post-production inspection and testing of products. If the product did not fit or work right, it was rejected or reworked. This clearly uneconomical approach has been replaced in many industries by a process control approach which directly results in better quality as well as improved productivity. An increasing number of standards are including quality assurance provisions which are clearly PPMs.[4]

Sometimes it may be technically difficult or even impossible to control product characteristics by inspecting the product. For example the ISO

9000 quality system uses the term 'special processes' to describe processes that cannot be fully verified by subsequent testing of the product, and where process deficiencies may become apparent only after the product is in use.[5] Plastic moulding products are a clear example. The ability of a moulded product to withstand stress may depend on the uniformity of dispersion of additives, the moulding temperature and other factors in the production process.[6]

PPMs addressing environmental effects not transmitted by the product

It should be noted that in a majority of cases, environmentally based PPMs will not affect the product's final characteristics, but are intended only to control environmental effects caused during production. As long as there are no transborder environmental effects, the environment of the importing country will not be affected by PPMs used in the exporting country. In these cases, protection of the importing country's domestic environment does not require that products imported into that country meet its own PPM standards.

PPMs and international trading rules

The 1979 Tokyo Agreement on Technical Barriers to Trade does not cover PPMs. Although the Uruguay Round's Technical Barriers to Trade (TBT) Agreement's technical standards and regulations do cover 'product characteristics and their related processes and production methods', the term is limited to coverage of PPMs that are characterised in the final product.

The WTO prohibits discrimination between 'like products' imported from different countries and between imported and 'like domestic products'.[7] The 'like product' issue has become relatively important in the context of trade and the environment. Recently, Austria announced a mandatory label programme for tropical timber and tropical timber products. It was intended that all commercial timber would bear a label displaying the inscription 'made from tropical timber' or 'contains tropical timber'. The ASEAN countries argued strongly that the label requirements were inconsistent with the provisions of the GATT, as the regulations applied only to tropical and not to other types of wood, which were 'like products'. The 'like product' issue was also very important to the decisions made in the tuna–dolphin cases.

There is no definition of 'like products' in the WTO and it has been up to dispute panels to determine what constitutes a 'like product' on a case-by-case basis, taking into account the objective and purposes of the regulation or measure. Dispute panels have generally taken the view that while tariff classification could be one of the criteria, the final determination should not be made solely on this basis. Other criteria, such as the nature of the product, its intended use, its commercial value, price and substitutability should also be taken into account when determining what is or is not a 'like product'.[8]

PPM-BASED STANDARDS AND THEIR IMPACT ON MARKET ACCESS AND COMPETITIVENESS

The differences between product and process regulations and their impact on competitiveness may at times be unclear for a number of reasons. For instance, product regulations may mandate the use of a specific production technology. Technological developments may change the conventional relationship between PPM standards and competitiveness.[9] Consumer preferences for 'green' products may require changes in PPMs,[10] and in the context of multilateral environmental agreements, the difference between the mandatory nature of product regulations and the voluntary nature of PPM-based regulations becomes less pronounced.

With regard to the competitiveness effects of PPM-based environmental standards, on the other hand, the following three issues are particularly important.

Compliance with domestic environmental regulations

The debate about the competitiveness effects of environmental regulations has focused on the costs of compliance and the variations between environmental regulatory systems in different countries. Although more stringent process standards and regulations may produce economic benefits in terms of improved human health and efficient use of resources, they may also adversely affect competitiveness at the sectoral or enterprise level. While the compliance costs at this level appear higher in a static analysis, a dynamic analysis shows lower costs since incentives for innovation and the use of 'clean technologies' may result in cost savings over the long term. In addition, some sectors will actually benefit if large markets exist for that sector's particular environmental goods and services.

Studies have shown that on average the environmental costs of production are only 1–2 per cent, although there can be a more significant impact on capital costs, particularly in pollution-intensive industries such as the petroleum industry.[11] In addition, most of these studies have considered only part of the environmental protection costs, such as specific industrial pollution control costs.[12] Moreover these studies have not picked up micro impacts. For instance, in particular sectors or firms, differences in process standards and regulations may have more serious cost impacts if environmental externalities are more fully internalised. For example carbon taxes may have a profound impact on international competitiveness.

The issue of whether 'dirty' industries migrate in response to differences in PPM regulations raises two concerns. First, the 'push effect' of stringent regulations can lead to an exodus of polluting industries, and second, the 'pull effect' of laxer environmental policies can lead to a 'pollution haven' effect.

While there is no conclusive proof that polluting industries from OECD countries have migrated to developing countries, in some sectors, such as asbestos, heavy metals and leather tanning, some relocation of dirty processes does seem to have occurred. Migrating industries are generally, however, those that are relatively less competitive and environmental regulations may be only one of the many reasons why they migrate. Arguments that differences in PPM-based standards lead to production migration and employment loss in OECD countries have not yet been verified.

Nonetheless studies do indicate that pollution-intensive industries are growing faster in developing than in developed countries. Whether this can be attributed, at least in part, to the migration of polluting industries because of differences in PPM standards remains inconclusive. What does seem clear, however, is that developing countries are unlikely to use lower standards as part of their 'strategic policy' to gain competitiveness.[13] This is because there will be an imposition of social costs in terms of health and safety risks for workers as well as for the inhabitants in and around factories. In addition, products manufactured by these industries may not find markets in countries with high environmental standards and these migrating industries are typically stagnant or declining.[14]

It should be noted that both government and industry are interested in shaping environmental policies in a way that removes strong negative impacts on competitiveness. Gradual introduction, exceptions, rebates and compensating subsidy schemes are used to mitigate the effects on international competitiveness and trade. By absorbing part of the environmental costs through subsidies, governments have often been able to mitigate the possible short-term negative impact of higher standards on the competitiveness of their firms. In some cases, by bearing part of the capital costs and financing research into environmentally sound technologies, governments may contribute to the dynamic adjustment of firms to higher standards.[15]

In expanding sectors, differences in environmental standards between countries should not create competitiveness problems. However in natural-resource-intensive sectors that are heavily dependent on price as a factor of competitiveness, even small differences in standards may have a significant impact.[16] This applies especially to low-value-added products, which constitute the bulk of the exports from developing countries. The ability of developing countries to implement higher environmental standards will also be limited by financial and technological constraints and they may find it difficult to subsidise the development of environmentally sound production methods.

Maintaining competitiveness in external markets

At times producers incur costs for environmental investments because of demands arising in export markets. For example it is believed that part of the investment costs incurred by the Canadian and Swedish pulp and paper industry is a result of consumer pressure in Europe, particularly Germany, for chlorine-free paper, and not solely of higher standards.[17]

Although PPM-related trade restrictions are not normally allowed under current international trade rules, in some cases they may play a role in overall consumer acceptance of a product. Both the product itself and the process-related environmental 'virtues' of the product may become factors of competitiveness in environmentally conscious consumer markets.

In some of the 'greener' European countries, for instance, 'product policies' used to control the environmental impacts of products sometimes also include a life-cycle assessment of their environmental impacts. Some eco-labelling programmes already apply PPM-related criteria. PPM-related issues are also sometimes considered by retailers and importers, and are used in government procurement guidelines, particularly by local governments. Certain voluntary codes or covenants include PPM-related policy measures that may have an impact on trade, such as the Netherlands Framework Agreement on Tropical Timber (NFATT).[18]

More research is needed to analyse the true extent to which the above-mentioned factors affect market access and competitiveness. Consumers in OECD countries may show a preference for specific PPMs favouring domestic products over imported ones. For example, if sustainable forest management is an important element in the choice of wooden furniture in OECD countries, then temperate wood products may automatically be substituted for tropical wood products. Often consumer preferences for specific PPMs may be related to familiarity and knowledge of the environmental characteristics of certain domestic processes, thus forcing developing countries that use different but equally environmentally friendly PPMs to obtain international certification to testify that their products are environmentally friendly. Otherwise they may be unable to capitalise on the environmental virtues of their products.

The competitiveness of firms in developing countries may be impaired by stringent environmental standards if they have to raise their standards to the same level as those of developed countries. An exporting firm may have to incur additional costs in order to comply with PPM-based regulations in the importing country; but on the other hand, if there is no comparable demand for such standards in its domestic market, the exporting firm will not be able to realise premiums and possible economies of scale over its entire range of production. Moreover subsidies may not be so readily available to these firms.

In any case, it is important that the impacts on trade and sustainable development in developing countries are taken into account. In certain cases, such as eco-labelling, the establishment of the 'Code of Good Practice for the Preparation, Adoption and Application of Standards' in the Uruguay Round agreements could make a contribution in this regard. For example, in accordance with Article 4 of the TBT Agreement, governments are to 'take such reasonable measures as may be available to them to ensure that local governments and non-governmental standardising bodies within their territories ... accept and comply with this Code of good practice'.

The subsidy question

Direct financial assistance provided by governments to producers in order to assist them in complying with environmental regulations is considered a subsidy by the WTO, with the WTO provisions on subsidies and countervailing measures being applied. In the Uruguay Agreement on Subsidies and Countervailing Measures some environmental subsidies have been made exempt from the general rule that 'specific' subsidies are 'actionable'. Assistance to firms for the adaption of their existing facilities to conform with new environmental regulations that result in greater financial burdens are 'non-actionable', provided certain other conditions are met.

Some subsidies eliminate external costs or capture environmental benefits, and these should not be actionable, in principle, under the WTO.[19] Developing countries may have special needs in the field of subsidies. However, because of competing financial priorities they often cannot provide compensating subsidies for environmental standards. International aid and technical assistance may be required if developing countries are to adhere to new environmental PPM standards.

PPMS AND HARMONISATION

It is sometimes suggested that harmonisation of PPMs could be a mechanism for alleviating competitiveness concerns as well as being beneficial to the environment. Environmentalists at times favour harmonisation of process standards as a guarantee against competing deregulation. From an environmental point of view, harmonisation guarantees a commitment to specified environmental objectives.[20] Some industry groups and certain labour groups have also favoured harmonisation as a means of combatting import competition and avoiding the migration of pollution-intensive industries. There are, however, several arguments against harmonisation, particularly because of the fact that assimilative capacities and social preferences vary between countries.

Is harmonisation desirable?

While harmonisation may be justified to address global environmental problems, countries nevertheless have different responsibilities for and means of addressing these problems. On the basis of equity, cases can be made that developing countries should receive transfers of funds and technology to enable them to implement internationally agreed environmental programmes.

When dealing with intrinsically domestic environmental problems, each country should be allowed to adopt standards that reflect its own environment and development conditions. Developing countries must build up an adequate framework of environmental rules and regulations that can be effectively enforced in order to encourage the switch to more environmentally friendly processes and technologies, and the orientation of technology transfers towards more environmentally sound technologies. The implementation of environmental standards may require careful adjustment, allowing for different levels of technological and socioeconomic development and taking into account the uniqueness of countries.[21]

Differences in process regulations may be appropriate. For instance the 'polluter pays principle', while encouraging harmonisation, does not require it. Indeed it recognises that differences in standards may be justified by a variety of factors. In certain cases harmonisation of process standards could mask comparative advantage.

It should also be noted that 'Uniform emissions standards across countries do not ensure that those countries will attain the same levels of ambient quality, nor do they ensure that all competing firms in those countries will face the same compliance costs'.[22]

The case for minimum standards

While harmonisation of standards could be encouraged where the same environmental and economic conditions prevail, according to some observers internationally agreed minimum standards could be useful in other cases. These types of standard may be a possible means of avoiding trade friction arising from differences in environmental standards.[23] There are, for example, codes of conduct that represent accepted minimum standards for air pollution emissions.

Where PPMs have no transborder environmental effects, compliance with minimum standards should be voluntary. Trade restrictions against non-complying imports would not be necessary from an environmental point of view since there would be no impact on the environment of other countries.

Internationally agreed minimum standards could also be linked to positive measures such as the transfer of finance and technology, and general environmental support. A case in point is the Montreal Protocol, where special and differential treatment was accorded to developing countries and a multilateral fund was established to meet their financial and technological requirements in order to encourage them to meet reduction targets.

Is eco-dumping an issue?

It has been argued that cost differences arising from variations in PPM standards between countries can act as trade-distortive subsidies and sources of 'unfair competition'. According to these views, international trade rules should be amended to permit the use of subsidies, countervailing duties or other measures to 'level the competitive playing field'.

'Eco-dumping' is a situation where a country deliberately sets its standards at an artificially low level or does not enforce them in order to gain a competitive trade advantage or attract investment. It has been suggested that environmental countervailing duties be levied on imported products in these cases. While such types of countervailing duty are not permitted under present WTO rules and have never been applied in practice, there have been several threats of their being applied. Environmental countervailing duties presume that foreign process standards should be comparable to the standards of the importing country.

Supporters of environmental countervailing duties argue that lax environmental regulations should be considered an implicit subsidy. In their view it is unfair to allow certain countries to apply lax environmental regulations or not to enforce them, while at the same time it is fair to provide governmental assistance to firms so that they may comply with more stringent environmental regulations. WTO does not recognise the concept of implicit subsidies.

It is widely believed that environmental countervailing duties are undesirable from both an economic and environmental viewpoint. An efficient international environmental policy by no means requires that production standards in different countries should be uniform. The use of environmental countervailing duties may contribute to trade friction and adversely affect the multilateral trading system.[24]

One of the propositions made in Agenda 21 is to 'Seek to avoid the use of trade restrictions or distortions as a means to offset differences in costs arising from differences in environmental standards and regulations, since their application could lead to trade distortions and increase protectionist tendencies'.[25] The adoption of reasonable environmental standards and wider adherence to the 'polluter pays principle' would also help avoid trade frictions over 'hidden' environmental subsidies and eco-dumping.[26]

TRADE RESTRICTIONS AND PPMS

Trade measures used for environmental purposes have tended to include both trade restrictions and positive trade measures. This section deals with trade restrictions while the next deals with positive measures, such as 'green' trade preferences.

Are PPM-related trade restrictions appropriate?

Firms or environmental groups in one country may object to PPMs used in other countries because they either do not internalise environmental costs or constitute 'unfair' competition. Alternatively, PPMs may be objected to because they create transborder or global environmental effects, such as global warming, or because they conflict with certain 'values'.

Trade restrictions could be appropriate if taken within the framework of a multilateral environmental agreement. But in these circumstances PPM-related trade measures may in practice be quite difficult to implement. For instance, in the case of the Montreal Protocol it has not been considered feasible to include products that are made with but do not contain CFCs. This has partly been due to the fact that there are problems with inspection of PPMs that are not detectable in the traded product, and partly to the accelerated phase-out of CFCs by the signatories to the Protocol, which meant that this trade measure was not necessary.

Other cases of trade restrictions that tend to be both unilateral and extraterritorial are difficult to justify.[27] It should be noted that the use of trade restrictions in the context of competitiveness has already been dealt with. When 'values' are involved in it may be particularly difficult to arrive at an international consensus. It has been mentioned that in these cases eco-labelling could at times be a viable alternative.[28]

PPM-RELATED MEASURES WITH POTENTIAL POSITIVE IMPACTS ON TRADE

Rather than resorting to trade restrictions, alternative mechanisms for addressing PPM-related issues, such as technology transfer and financial and technical assistance, should be sought through international cooperation. Eco-labelling and environmental certification pursued at an international level may also be a viable alternative to trade restrictions and can help developing countries capture the rents associated with environmental concerns in industrialised countries.[29] At the same time such measures could enable developing countries to upgrade their PPM-based standards.

Financial assistance and transfer of ESTs

The need for the international community to provide funds for the so-called 'incremental costs of technology switching' to address global environmental problems has already been recognised with the creation of the Global Environmental Facility and the Multilateral Fund under the Montreal Protocol. However there are no comparable international mechanisms for the facilitation of the transfer of environmentally sound technologies (ESTs) that are appropriate for local environmental problems. In issues such as poverty alleviation, where the concerns are simultaneously of an environmental and a developmental nature, the transfer of ESTs urgently requires facilitating mechanisms.[30] Innovative financing mechanisms could include tradeable permits, international carbon offsets and incentives for green foreign direct investment.[31]

Measures that induce the trade of environmental protection goods, services and technology should also contribute to the upgrading of environmental standards in developing countries and will meld commercial and foreign policy environmental objectives.[32]

'Premiums' for PPMs

Under certain conditions, green consumerism in OECD countries could also help developing countries to upgrade their PPM standards. Premiums can provide a means of recovering the higher costs of producing a product in an environmentally friendly way, compared with using an old technique.

International cooperation on eco-labelling and certification may also be useful in meeting this challenge. In principle, the establishment of internationally agreed guidelines for environmental certification and the mutual recognition of national systems may be further possibilities. A number of options concerning eco-labelling are mentioned below.

'Green' trade preferences

According to some observers it is likely that OECD countries will in the future link further preferential access to their markets with environmental performance. Options are currently being explored for ways of granting trade preferences to 'green' products. 'Green' trade preferences will benefit developing countries only if they provide *additional* preferences. Environmental protection should not become a conditionality for existing preferences.

Eco-labelling

The purpose of eco-labelling is to promote the consumption and production of more environmentally 'friendly' products by providing consumers with information about their relative environmental impacts, based on a life-cycle analysis.

Few export products from developing countries are covered by eco-labels and to date the effect has been small. However the European Union is in the process of establishing eco-labels for products such as textiles, clothing, footwear and tropical timber. It should be noted that products such as textiles and footwear tend to have important 'upstream' environmental impacts, such as the water pollution caused by leather tanning industries, which may be as or more important than those at the consumption or disposal stage. Therefore PPM-related criteria may become relatively more important in the future.

PPM-related criteria that are based on environmental and technological conditions and priorities of the OECD countries may be difficult for developing countries to meet or be environmentally inappropriate. Possibilities for increasing the compatibility between eco-labelling and the trade and sustainable development interests of developing countries have been examined by UNCTAD's *ad hoc* Working Group on Trade, Environment and Development, for example through a discussion of concepts such as equivalencies and mutual recognition.[33]

The concept of equivalencies in the context of eco-labelling implies that when comparable environmental objectives can be achieved in different ways, taking into account the specific environmental conditions of each country, different criteria can be accepted as a basis for awarding eco-labels. The concept could be applied in two different circumstances. Firstly, the eco-labelling programme of the importing country may accept compliance with certain environmental requirements or the achievement of certain environmental improvements in the exporting country as 'equivalent' to compliance with specific criteria and thresholds established in its own programme, even when no eco-labelling programme exists in the exporting country. Secondly, the concept of 'equivalent' standards is generally considered to be a basic condition for mutual recognition of eco-labelling programmes.

Where environmental aspects are addressed through regulatory approaches, compliance with the exporting country's domestic regulation could be considered to be equivalent to compliance with the regulation in the importing country. Where non-regulatory PPM-related criteria are used to define environmentally superior products, for specific process-related criteria addressing intrinsically local environmental problems in the producing country, the eco-labelling programme of the importing country may be accepted as equivalent to PPMs that are environmentally friendly to the domestic environment of the producing country, taking into account its own environmental and developmental conditions.

The basic idea of mutual recognition is to recognise the validity of divergent environmental criteria and also to ensure that trade interests are not unduly affected by this diversity. The interest in mutual recognition of eco-labelling has been growing, partly as a result of concern that the emergence of different eco-labelling programmes in an increasing number of countries may adversely affect trade and create confusion among consumers. Mutual recognition, however, is not yet widely accepted. Environmental groups may be concerned that mutual recognition could imply that products that do not meet the same stringent criteria of the domestic programme are nevertheless awarded the corresponding eco-label. Domestic producers may be concerned about possible effects on competitiveness. Mutual recognition requires a confidence-building process. Acceptance by consumers and environmental interest groups requires the exporting country's programme to be credible.

Mutual recognition in the context of eco-labelling generally would imply that, provided that certain conditions are met, qualification for the eco-label of the exporting country is accepted as a basis for awarding the eco-label used in the importing country. Mutual recognition would normally apply to identical or similar product categories.

Reciprocity or mutual recognition, based on some type of international agreement, could take several forms.[34] The eco-labelling programme of the importing country could, for example, agree to award its own eco-label to products that (a) meet the criteria of the eco-label of the exporting country; or (b) meet the PPM-related criteria of the eco-label of the exporting country, and the use and disposal criteria of the eco-label of the importing country (as certified by the exporting country's programme).

The few proposals for mutual recognition of eco-labelling schemes that have been formally discussed so far involve, on the one hand, the European Union and the EFTA countries, and on the other the United States and Canada, although these have been limited to certain aspects of eco-labelling such as conformity assessment procedures. Mutual recognition of eco-labelling programmes implemented by countries at different levels of economic development may involve programmes that vary more substantially from each other in terms of environmental criteria. Mutual confidence, based on the previous harmonisation of technical requirements, such as testing and inspection methods, would be a prerequisite for mutual recognition. Work on internationally agreed guidelines for eco-labelling, for example in the ISO, could also contribute to creating conditions for moving towards mutual recognition of eco-labels.

Certification of environmentally friendly products

Three suggestions are being made to enable countries to improve their environment and facilitate trade between nations. These include principles for the mutual recognition of environmental certification, equivalencies between environmental certification authorities and internationally agreed guidelines for certification.

Principles for the mutual recognition of environmental certification could be relatively easily established and have already been discussed in the context of eco-labelling.

Equivalencies could also be established through specific process-related criteria and thresholds. This would be particularly appropriate for processes that are considered environmentally friendly to the domestic environment of a producing country, taking into account its own environmental and developmental conditions.

Internationally agreed guidelines for the certification of environmentally friendly products could also be developed. While broad guidelines could be negotiated at the international level, the formulation of more specific criteria, testing and monitoring should be left to national standardisation bodies. These authorities could also authorise the use of an internationally agreed mark or symbol indicating the environmentally friendly qualities of the product after confirming that they had abided by the internationally agreed guidelines. Training and other assistance on testing procedures could also be provided to developing countries. This system of certification could be based on a comparison of the production processes and products within an individual country. The system could also be applied to similar products produced by different production processes in different countries, on the understanding that only certain production processes are feasible in certain countries, for instance because of particular natural circumstances.

While testing and certification would be handled by national authorities, checks and balances for the system would be best left to the markets. The fiercely competitive nature of the export markets of developing countries suggests that if symbols and marks were to be unjustifiably used, consumer and producer complaints would be a strong deterrent to the continued use of international certification. However, in cases where these complaints could not be resolved, an investigation team could be set up under the aegis of the ISO to conduct on-site plant testing.

Sustainable production of commodities

The introduction of sustainable practices for the production of internationally traded commodities may require innovative forms of cooperation between producers and consumers. In fact studies are already underway regarding the possibility of facilitating the internalisation of environmental externalities in the production of commodities with the help of international commodity-related environmental agreements (ICREAs). Various options are being analysed to determine how financial transfers or environmental premiums could be provided, and whether or not they would need to be combined with preferential market access.

CONCLUSIONS AND RECOMMENDATIONS

PPM-related measures are an important instrument of environmental policy making. But trade restrictions may not be appropriate or effective in dealing with PPM-based environmental problems. The efforts of developing countries to internalise environmental costs, through PPM-based regulations as well as other mechanisms, must be encouraged and receive positive international support. International cooperation must be based on the statements made in the Rio Declaration and Agenda 21, and aim at accelerating development, maintaining an open trading system, improving market access and building institutional capacity to integrate trade and environment policies within the framework of national policies for sustainable development. Specific measures for achieving all of this include transfers of technology, financial and technical assistance, the expansion of market opportunities for 'green' products, and certification of environmentally friendly products.

Although the use of trade restrictions, based on both product and PPM-related criteria, may be permissible in the framework of a multilateral environmental agreement where global environmental objectives are being promoted, and where there are transborder or global environmental impacts, no strong economic or environmental grounds exist for the imposition of trade restrictions on products that do not comply with the domestic PPM standards of the importing country.

In addition, because the cost effects of PPM-based standards are not very significant and have not led to the significant migration of dirty industries, the use of trade restrictions to offset differences in PPM-based standards does not appear to be justified on either environmental or competitiveness grounds. The risk of abuse in any particular use of trade measures will outweigh any competitive benefits that may be obtained.[35]

As mentioned in Agenda 21, measures to offset differences in costs arising from differences in environmental standards and regulations may increase protectionist tendencies and may also lead to trade friction. Imposing certain PPMs on developing countries will tend to be economically inefficient and will seriously affect their developmental progress.

Sometimes consumer preferences may affect market access conditions and the competitiveness of specific products based on the PPM used in their production. These forces have the potential to work either as an obstacle to trade or as a means of helping to improve PPMs through trade, by providing premiums to environmentally friendly products. Measures that have a certain degree of government involvement, such as eco-labelling systems and certain covenants, should henceforth be reported to the WTO. Wide adherence by local and non-governmental bodies to the new code of good practices also appears to be essential.

Efforts by developing country exporters to obtain environmental premiums for environmentally friendly products in OECD markets should be encouraged. This could best be achieved when initiatives involving products of export interest to developing countries are based on an international process that has their fullest possible participation.

Certification schemes that take account of environmental and trade objectives in the context of PPMs could be based on three elements. First, certification should recognise the concept of equivalence, so that environmental improvements of utmost importance to the producing countries are rewarded with certification. Second, principles and guidelines could be developed for the mutual recognition of these certification schemes, such as eco-labelling, taking into account the environmental and developmental priorities of developing countries. Finally, internationally agreed guidelines could be used to provide an outline of broad criteria for certification, the formulation of specific standards, testing and monitoring being left to local standardisation bodies.

Notes and references

* The views expressed in this chapter are those of the authors and do not necessarily reflect those of the United Nations Conference on Trade and Development.

1. Normally the term regulation is used when compliance is mandatory, while the term standard is used when compliance is voluntary.
2. See OECD, *Conceptual Framework for PPM Measures* (Paris: OECD, 8–10 March 1994).
3. For the sake of analytical clarity, PPMs that are 'related to a product', that is, affect the characteristics of the product, are considered under product regulations. Process regulations are thus assumed to cover only PPMs that are not related to the product, such as emission standards for steel plants.
4. GATT, 'Process and Production Methods – Draft Proposal by the United States', TBT/W/108, 23 February 1988, p. 2.
5. See International Trade Centre UNCTAD/GATT, *ISO 9000 Quality Management Systems: Guidelines for Enterprises in Developing Countries* (Geneva: ITC, 1993), p. xvi.
6. See ibid.
7. See Articles 1 and 3, General Agreement on Tariffs and Trade, and Article 2.1 of the Technical Barriers to Trade Agreement.
8. Vinod Rege, 'GATT Law and Environment Related Issues Affecting Trade of Developing Countries', *Journal of World Trade*, vol. 28, no. 3 (June 1994).
9. UNCTAD Secretariat/Institute for Economic Analysis, 'The Role of Technology in Environmentally-Motivated Structural Change and the

Implications for International Trade', paper prepared for UNCTAD under the UNCTAD/UNDP project on Reconciliation of Environmental and Trade Policies, December 1993.

10. Ibid.
11. Charles Pearson, 'Trade and Environment: The United States Experience', paper prepared for UNCTAD under the UNCTAD/UNDP project on Reconciliation of Environmental and Trade Policies, December 1993.
12. One shortcoming of many studies is that they focus almost exclusively on industrial pollution-control costs. For a summary of the limitations of different studies see Congress of the United States, Office of Technology Assessment (OTA), *Trade and Environment, Conflicts and Opportunities*, Appendix E. OTA-BP-ITE-94 (Washington, DC: U.S. Government Printing Office, May 1992).
13. See S. Barrett, 'Strategic Environmental Policy and International Competitiveness', paper prepared for a workshop on Environmental Policies and Industrial Competitiveness, OECD, Paris, 28–29 January 1993.
14. Lyuba Zarsky, *Trade-Environment Linkages and Sustainable Development*, report of the Department of Arts, Sports, Environment, Tourism and Territories, Nautilus Pacific Research, North Fitzroy, Australia, October 1991.
15. Piritta Sorsa, 'Competitiveness and Environmental Standards: Some Exploratory Results', paper presented to the Conference on International Trade and the Environment, Austrian Federal Economic Chamber, 22–23 March 1993.
16. See H. Verbruggen, 'The Trade Effects of Economic Instruments', paper prepared for the OECD Environment Directorate Informal Experts Workshop on Environmental Policies and Industry Competitiveness, 28–29 June 1993.
17. See US Congress, Office of Technology Assessment, 'Industry, Technology, and the Environment – Competitive Challenges and Business Opportunities', OTA-ITE-586 (Washington, DC: US Government Printing Office, January 1994), p. 199.
18. See 'The Case of Tropical Timber in the Netherlands', paper submitted by the Netherlands delegation to the high-level session of the OECD Environmental Policy Committee, Paris, 7–8 December 1993.
19. C. S. Pearson and R. Repetto, 'Reconciling Trade and Environment: the Next Steps', paper prepared for the Trade and Environment Committee of the US Environmental Protection Agency, 1992.
20. See, for example, Latin American Economic System (SELA), 'Trade, Environment and the Developing Countries', (SP/LC/XVIII.O/Di No.2).
21. UNCTAD and the Government of Norway, 'Report of the Workshop on the Transfer and Development of Environmentally Sound Technologies', Oslo, 13–15 October 1993, p. 22 (Geneva: UNCTAD, 1993).
22. See Robert Repetto, 'Trade and Environment Policies, Achieving Complementarities and Avoiding Conflicts, Issues and Ideas' (Washington, DC: World Resources Institute, July 1993).
23. Latin American Economic System (SELA), op. cit.
24. The GATT Report on Trade and the Environment has very strongly opposed the use of trade measures such as countervailing duties to offset the compet-

itive effects of differences in standards across countries: 'To allow each contracting party unilaterally to impose special duties against whatever it objects to among the domestic policies of other contracting parties would risk an eventual descent into chaotic trade conditions similar to those that plagued the 1930s'. GATT, 'Trade and the Environment', in *International Trade 90–91*, vol. 1 (Geneva: GATT, 1992), p. 20.

25. Paragraph 2.22(e), Report of the United Nations Conference on Environment and Development, Annex II, June 14, 1992, UN Doc. A/CONF.151/26 (1992).
26. Robert Repetto, 1993, op. cit.
27. The Business and Industry Advisory Committee of the OECD(BIAC) has observed that 'Unilateral trade measures imposed on the basis of the process or production method used to make a product are unacceptable. Such measures assume that importing countries have the right to pass judgements on the domestic policies of their trading partners, and to impose their judgements through trade instruments. There is no limit to the extent to which such unilateral measures could be used because no two countries have (nor should they necessarily be expected to have) equivalent environmental policies and standards in all areas. The use of trade measures in such circumstances would tend to lead to their imposition for any difference between national environmental policies and standards (not to mention policies in the areas of labour, tax, competition law, etc.) and to rapid decline in international law'. Business and Industry Advisory Committee of the OECD, 'BIAC Statement to the Special Session of the OECD Environmental Policy Committee' (Paris: OECD, 7–8 December 1993), p. 3.
28. Pearson, 1993, op. cit., p. 17.
29. Panayotis N. Varangis, C. A. Primo Braga and Kenji Takeuchi, 'Tropical Timber Trade Policies: What Impact will Eco-Labelling have?', paper prepared for the UNCTAD seminar on Ecolabelling and International Trade, 10 May 1993.
30. UNCTAD and the Government of Norway, op. cit.
31. See 'Financial Resources and Sustainable Development: International Policy Options', discussion paper prepared by the UNCTAD Secretariat for the Preparatory Meeting on Finance, Kuala Lumpur, 2–4 February, 1994.
32. For example, in the United States the Overseas Private Investment Corporation has proposed an environmental investment fund to stimulate environmental investment in developing countries. The Energy Policy Act of 1992 directs the secretary of energy, through AID, 'to create a technology transfer program aimed at reducing the U.S. trade deficit through the export of innovative environmental technologies'. The Export Enhancement Act of 1992 seeks to encourage exports of environmental goods and services. See Pearson, 1993, op. cit.
33. See: 'Eco-labelling and market opportunities for environmentally friendly products', report by the UNCTAD secretariat (TD/B/WG.6/2).
34. Environmental Choice Program, Environment Canada, 'Dealing with the trade barrier issue', paper presented to the UNCTAD workshop on eco-labelling and trade, Geneva, 28–29 June 1994.
35. See Repetto, 1993, op. cit.

3 Environmentally Orientated Product Policies, Competitiveness and Market Access

Veena Jha and René Vossenaar*

Environmentally orientated product policies establish certain criteria for the design, content and/or disposal of products in order to minimise their environmental impact. They often take the form of standards and regulations, market-based instruments, product liability requirements or labelling and information conditions. These policies may be voluntary in nature or mandated by governments.

Product specifications and the way they are formulated and applied may act as non-tariff barriers to trade. This can occur, for example, if potential exporters are discouraged by the complexity of the specifications or there is a lack of transparency concerning their application. Procedural delays and poor administrative practices may also magnify these problems.

This chapter examines the trade effects of environmentally orientated product-specific policies with special reference to packaging and eco-labelling. It analyses the impact of these policies on market access and competitiveness, particularly for developing countries.

EMERGING PRODUCT-SPECIFIC POLICIES

Although environmental policy making in most countries remains focused on the environmental effects of industrial activity, concern about the impacts of the products themselves has led to the formulation of policies aimed directly at the products.[1] Features of these newly emerging environmental policies are likely to have an important impact on market access and competitiveness. Such features as their voluntary nature, their technical characteristics, the use of process-based product characteristics, and the somewhat nebulous issue of the 'environmental quality' of the product all pose significant potential barriers to trade. It is also important to note that these features of environmentally based product policies distinguish them from other performance-based product policies. They therefore merit special attention.

The environmental problems targeted by product-related policies include the improvement of material and energy efficiency, waste management and the control of hazardous or environmentally harmful substances. Among the most commonly used policies are labelling and other information-based instruments, economic instruments and producer liability.

Environmental problems addressed by product-specific policies

A range of different policies are used to attend to environmental problems caused by products. For instance in some countries, such as the United States, where there is widespread public concern about the environment, bans on the use of environmentally hazardous substances may be easier to accomplish than the setting of technical standards that involve complex risk assessment. However outright bans raise concerns about market access for products exported from other countries. A case in point is the ban invoked by Germany on the use of PCPs in the manufacture of leather goods. Until other substitute chemicals started to be used, the exportation of leather products from India to Germany declined significantly.

Setting limits on the concentration of hazardous substances is a more common method of regulation than outright bans. Meeting these limits may, however, be difficult for firms in developing countries. Gaining access to precise measurement technology and developing credible certification procedures may prove to be expensive and difficult, thus impacting on the competitiveness of their export products.

Apart from hazardous substances, environmentally orientated product policies are also being used to reduce energy consumption. For example the US National Appliance Energy Conservation Act sets federal energy standards for household appliances such as refrigerators, dishwashers and television sets. Many other OECD countries are also considering implementing energy standards for products and energy efficiency policies for production processes.

Energy standards can be implemented variously. For instance products that do not meet a standard may be denied access to markets, or may need to be labelled so that the consumer can determine a product's energy use, again affecting their competitiveness.

In the area of waste management, product policies aim to reduce the production of waste at source and recycle or reuse any waste produced. Policies have tended to focus on recycling rather than reducing or reusing waste. Recycling policies not only require the promotion of the recyclability of products, but also the creation of a market for recycled materials.

Environmental regulations can require that products contain a minimum amount of recycled materials, thus creating a market for those products. An example of this is the state of California's recycled content requirement for newsprint and its effect on newsprint manufactured in and exported to the United States from Canada. In order to meet this requirement, firms were forced to import waste newsprint from the United States, mix it with virgin Canadian newsprint and then reexport the mixed product back to the United States. It is doubtful that this particular strategy benefited the Canadian or the US environment, or improved the competitiveness of the Canadian newsprint industry.[2]

Instruments used to implement product-specific policies

Instruments are now being used that seek to influence rather than regulate products and production processes. Of particular note are labelling mechanisms and economic instruments such as fees and taxes, as well as proposals to make producers responsible for their products throughout the product's life-cycle.

In order to reduce their own impact on the environment, both manufacturers and consumers require information about the relative environmental impacts of the various products they purchase. In the case of manufacturers, the requirement to provide environmental impact information about a product may of itself induce improvements in environmental performance in order to remain competitive.

Product labelling for environmental purposes may include negative or positive information and can relate to any particular environmental concern associated with the product during its life-cycle. This will involve additional costs for research and certification systems, further impacting on the competitiveness of products from developing countries, where firms, especially smaller ones, will find these costs onerous.

Economic instruments include taxes, product charges, levies, returnable deposit systems and various forms of penalty. These affect the options open to a producer and, in the environmental context, influence their behaviour in a way that is environmentally favourable. They may also generate revenue from a polluter that can be used by a community to mitigate environmental degradation. Returnable deposit systems are a common example. A consumer makes a deposit that is returned when the packaging item, such as a glass soft-drink or beer bottle, is returned to the purchase outlet.

Economic instruments can have significant effects on the market access and competitiveness of a product, depending in part upon the price elasticity, ease of substitution and income elasticity of the product. For example a border tax imposed on tropical timber products could result in export losses because it is easy to substitute temperate timber. Similarly, returnable deposit systems increase the costs of production by a higher margin for exporters than they do for domestic producers. In fact, while economic instruments can distort developing countries' trade and that of their exporters, concomitant environmental improvements in these same countries are doubtful. There are also questions about the efficacy of using these types of instrument in undeveloped markets. For example, rising inflation in almost all Eastern European countries means that it is cheaper for firms to pay fines or fees than to invest in pollution-abatement technology.[3]

An important emerging environmentally orientated product policy, and one that may have serious trade effects, is the proposal that producers should be responsible for their product throughout its entire life-cycle. This could require producers to take back their products after use. For products that enter into international trade, this is obviously not a realistic option and is likely to make internationally traded products uncompetitive. An alternative approach is to impose legal liability on the producer for products that damage the environment. This approach is presently being examined in several OECD countries, particularly in the United States.

PACKAGING

Packaging and the mitigation of its possible trade effect is a rapidly developing area of environmental policy making.[4] A number of countries have implemented policies that set targets for reducing packaging waste. Policies can provide for mandatory recycling systems or cover only voluntary industry agreements. They may also focus on packaging for specific purposes, such as beverage containers, or on specific substances, such as PVC. The following section will assess the possible impacts of packaging policies on market access and export competitiveness.

Administrative procedures in the area of packaging mean that producers and importers may incur increased costs. Since the costs involved in reuse or recycling depend on the size, quality and characteristics of the packaging, these costs may be higher for imported products than they are for similar domestic products. In addition, low-value packaging materials commonly used by firms in developing countries may suffer.[5]

Compliance costs are incurred when a producer changes the design or materials of the packaging in order to comply with the environmental regulations of the importing country. Eliminating hazardous substances or making materials recyclable can be costly. Measures used to create a market for recycled packaging materials may oblige exporters to import used packaging or used materials. Compliance with recycled content requirements may be particularly difficult for developing countries as their waste management programmes tend to be undeveloped.

Differing packaging regulations between countries, especially between those within the same geographical area, may create significant transaction costs. For example, in Germany incineration for energy recovery is permitted only if reuse or recycling are technically or economically unfeasible, while it is considered an acceptable alternative in France, Sweden and Japan. Harmonisation of countries' packaging requirements is desirable from a trade and an efficiency point of view. A recently drafted EU directive on packaging will permit member states to adopt higher targets than those laid down in the directive, provided that distortions to the EU's internal market are avoided and other member states are not hindered in their efforts to achieve their own targets. The difficulties involved in harmonising packaging regulations are highlighted by recent requests from a number of countries for derogation from the directive.[6]

The collection of packaging waste has often exceeded domestic recycling facilities and caused profound market disruptions in third countries. For example it appears that the German Packaging Ordinance has provided a *de facto* subsidy to German producers of recycled paper, who have benefited enormously from the substantially reduced paper costs that are a side-effect of the ordinance. This has had an enormous detrimental impact on the manufacturers of recycled paper in other countries and on the manufacturers and exporters of virgin pulp and paper.

In addition, concerns exist that exports of waste as secondary raw material for recycling or recovery, especially to developing countries, may actually be concealing the export of these wastes for disposal. Furthermore, it appears that imports of waste paper into Colombia, originating in the United States, Venezuela and certain Central American countries, may actually be discouraging the recovery of domestic waste in Colombia.[7] This is of concern because in many developing countries, waste recovery activities may be one of the only sources of revenue for the very poor.[8] However, this issue raises problems that go well beyond concerns about policies in the area of packaging.

New packaging policies may also induce exporters to substitute less environmentally friendly materials to suit importers' existing recycling facilities. The preference for easily recyclable materials may also create obstacles to packaging that uses a mixture of different materials. For example problems have arisen with the use of jute for wrapping bales of cotton and wool because of the steel clips that are used to secure them. Recycling companies in Germany have refused to handle these materials because of the labour required to remove the clips. Consequently textile importers have advised their suppliers to switch from using jute wraps to polyethylene wraps.[9] Coffee exporters, too, have been facing similar problems and have begun to convert to synthetic wraps.

ECO-LABELLING

The purpose of eco-labelling is to promote the consumption and production of more environmentally friendly products by providing consumers with information about the relative environmental impacts of similar products. In order to analyse the impacts eco-labelling may have on export competitiveness and market access, it is important to note that eco-labelling promotes product differentiation on the basis of environmental quality.

In the past, eco-labelling was less important in terms of international trade, in particular for developing countries. Recently, however, eco-labelling has become more important for developing countries. For example the European Union is in the process of establishing eco-labels for textiles, clothing, tropical timber and footwear. Approximately 45 per cent of the extra EU-imports in these products originate from developing countries.

Eco-labelling will primarily affect export competitiveness. However it may also affect market access when the eco-label is highly visible in a given product category, but is difficult to obtain for foreign producers. Eco-labelling can in some senses have the same effect as brand names. This will affect the market access of exporters from developing countries because of costly reputation building exercises or the costs of information gathering. In addition, the selection of new product categories, the determination of criteria and thresholds for labelled products as well as product testing and plant inspection, may favour domestic producers over foreign suppliers and act as an obstacle to trade.[10]

Preferences for eco-labelled products, or market shares, will depend on product category, producer response to eco-labelling, and the strategies adopted by the eco-labelling authorities. Programmes have proven difficult to implement and environmental labels in a number of product categories have largely been ignored by manufacturers.[11] In product categories such as paper or pulp, eco-labelled products account for a large share of the total paper and pulp sold in European markets.

Market shares also depend on the strategies adopted by individual eco-labelling authorities, and these strategies often vary considerably among different systems. For example the Canadian system set the thresholds for different criteria at high levels, so that initially only about 20 per cent of all products could qualify for the Canadian label. In contrast, in the case of the EU eco-label – the European flower – the strategy is to make it more visible from the outset so as to improve its acceptability in the market. It is difficult to determine which of these two strategies will be more beneficial to traded products. While the Canadian system initially leaves open 80 per cent of the market to unlabelled products, it is more difficult for an environmentally conscious exporter to obtain the label. On the other hand, because the European label is more visible, it may almost mandate the use of the label by exporters.

Evidence from Germany's Blue Angel and Singapore's Green Label systems confirms that consumers have been very responsive to both eco-labelling schemes.[12] Preliminary market surveys indicate that for certain categories of paper, batteries and detergents, the share of eco-labelled products in the total sales of product categories is between approximately 40 and 60 per cent.[13] Moreover a significant proportion of all consumers in Germany recognise the label, associate it with environmental friendliness and are willing to pay a premium for eco-labelled products.

Developing countries will become increasingly exposed to the effects of eco-labelling because criteria may increasingly focus on process and production methods (PPMs). For example products such as footwear and clothing tend to have 'upstream' environmental impacts that will be equally or more important than the impacts that occur at the consumption or disposal stage. Furthermore, new product categories will tend to include a larger share of products of export interest to developing countries. The costs of compliance with proposed PPM criteria may be very high, especially for small firms from developing countries. A case study concerning textiles in Turkey suggests that higher capital costs would not result in any significant reduction in operational costs, for example because the draft eco-labelling criteria of the EU do not induce firms in Turkey to use less material and energy or adopt other cost-saving practices.[14]

CONCLUSIONS

Environmentally orientated product-specific policies, emerging as they are in an international context, may have significant impacts on the market access conditions and competitiveness of exports from developing countries. These impacts include increased transaction costs for different product policies in different countries, increases in risk due to producer liability policies, and process or design modifications.

The costs of compliance will be higher for small firms than for large firms, especially if adjustments have to be made to the design of products. Furthermore it is doubtful whether developing countries will be able to obtain significant price premiums for making environmental improvements in the product sectors they are competitive in.

In some cases, product policies may have unintended trade and even environmental effects on trading partners. Provisions for reuse and recycling may represent a trade barrier, if not a *de facto* exclusion of the use of certain materials. The availability of recycling facilities in developing countries may affect the choice of materials.

Where life-cycle analysis is undertaken, certain PPM-related requirements may reflect the environmental priorities prevailing in the importing country but be inappropriate in the producing countries. Product profile information and eco-labelling are product policies that may particularly reflect local social and environmental values that are not shared by exporting countries.

Voluntary measures and self regulation are problematic in developing countries due to the doubtful credibility of testing and certification procedures. Despite the limitations on financial resources, improving these systems would no doubt yield substantial environmental and trade benefits. Sometimes it may be more cost effective to develop certification and eco-labelling mechanisms on a regional basis, and this possibility needs to be explored by developing countries.

Packaging requirements also affect export competitiveness and market access through administrative and compliance costs. Requirements may create uncertainty as to what materials will be acceptable to importers. Exporters need precise information about packaging requirements in the importing country and the existence of differences in requirements between countries. There is a need for better transparency and more technical assistance.

New and innovative policies will be required to reduce the adverse environmental impacts of products and to eliminate unsustainable forms of consumption in developed countries. One of the objectives identified in Chapter 4 of Agenda 21 is to 'promote patterns of consumption and production that reduce environmental stress and will meet the basic needs of humanity'.[15] Policies that promote these objectives have tended to be based on voluntary measures or on recently developed measures for which international trade rules are not yet well established. However the widespread application of these environmental measures is having an impact on international trade, and particularly on products traded from developing countries. Consequently there is a serious need to better regulate the use of these environmental measures and reduce their negative trade effects.

Notes and references

* The views expressed in this chapter are those of the authors and do not necessarily reflect those of the United Nations Conference on Trade and Development.

1. See G. Bennet and B. Verhoeve, 'Environmental Product Standards in Western Europe, the U.S. and Japan: A Guidebook', final report prepared under contract to the European Bank for Reconstruction and Development in cooperation with the Commission of the European Communities and USAID, December 1993.
2. See J. Grimmett, 'The Case of Recycled Paper in Newsprint', paper presented at an informal OECD expert workshop on Trade and Environment: Issues Pertaining to Processes and Production Methods (PPMs), Helsinki, 6–7 April 1994.
3. See G. Bennet and B. Verhoeve, op. cit.
4. See UNCTAD, *UNCTAD's Contribution, within its Mandate, to Sustainable Development: Trade and Environment – Trends in the Field of Trade and Environment in the Framework of International Cooperation* (Geneva: UNCTAD (TD/B/40(1)/6), August 1993).
5. S. Zarrilli, 'Ecopackaging Initiatives: Impact on International Trade and the Special Status of the Developing Countries', paper presented at the UNCTAD/SELA/ECLAC Regional Seminar on Environmental Policies and Market Access, Bogota, Colombia, 19–20 October 1993.
6. Greece, Ireland and Portugal consider that the recycling targets are too high. On the other hand Denmark, Germany and the Netherlands consider that the targets are too low. See *Agence Europe*, 16 December 1993.
7. See D. Gaviria, R. Gorney, L. Ho and A. Soto, 'Reconciliation of Trade and Environment Policies: The Case Study of Colombia', report prepared for the UNCTAD/UNDP project INT/92/207 (1994).
8. Anon, 'All that Remains: A Survey of Waste and Environment', *The Economist*, 29 May 1993.
9. See N. Robson, 'Jute Packaging and the Environment: The Problems and the Opportunities', paper prepared for the International Consultation on Jute and the Environment, The Hague, Netherlands, 26–29 October 1993.

10. V. Jha, R. Vossenaar and S. Zarrilli 'Eco-labelling and International Trade', UNCTAD Discussion Paper no. 70 (Geneva: UNCTAD, October 1993).
11. Ibid., p. 4.
12. See Chapter 4 and Jha, Vossenaar and Zarrilli, 1993, op. cit.
13. Ibid.
14. Celik Arouba, 'Analysis of Probable Impact of EU Eco-labelling Program and Related Criteria on Turkish Textiles and Garments Exports to European Markets', paper presented for the UNCTAD workshop on Eco-labelling and International Trade, Geneva, 28–29 June 1994.
15. See paragraph 4.7(a), Agenda 21, Report of the United Nations Conference on Environment and Development, Annex II, June 14, 1992, UN Doc A/CONF.151/26 (1992).

4 Environmental Policy Making, Eco-Labelling and Eco-Packaging in Germany and its Impact on Developing Countries

Christine Wyatt

Apart from the difficulties associated with unification, protection of the environment is considered by large sections of the German public to be the most important political issue in Germany today. Concern for the environment has also become a strong economic force, with the readiness of consumers to purchase environmentally sounder products resulting in steady sales growth for 'green' products. The attention paid by the German public to environmental protection is also an important concern for exporters. Many regulatory changes are taking place in response to public pressure for environmental protection and these are having a considerable impact on products imported into Germany.

This chapter discusses the evolution of these regulations in Germany, as well as a number of voluntary environmental measures that carry consequences for exporters. First, a brief overview of environmental policy making in Germany is given, along with trends in current regulatory, economic and voluntary environmental measures. The Blue Angel eco-labelling scheme is examined next, as well as a recent eco-labelling initiative in the textile sector. Both of these schemes have implications for the process and production methods used by producers in other countries. Probably of most concern for exporters is the recent Packaging Ordinance, which obliges producers to take back all product packaging. The efficacy of this ordinance, as well as the Dual System and Green Dot schemes established to comply with the ordinance, will be scrutinised. Of note for exporters to Germany is a discussion of the impact of the recycling of transport-packaging waste.

These new standards raise serious questions about their impact on trade. Although the new measures may pose barriers to trade, there is also the possibility of gains to be made for those exporters who can demonstrate the 'green' qualities of their products.

ENVIRONMENTAL POLICY MAKING IN GERMANY

Germany's political structure is characterised by the relative decentralisation of powers in domestic policy making. In the environmental domain, the federal and 16 *Länder* or state governments share legislative powers and management over various resources and environmentally sensitive activities.

A range of different legislative and policy instruments are used by both federal and *Länder* governments in achieving their environmental goals, including direct regulations, economic instruments and voluntary agreements. Although direct regulations have been the favoured policy instrument in the past, a growing consensus has emerged that the role of government in environmental policy making should be restricted to the establishment of broad legal frameworks. Consequently the use of economic instruments such as charges, deposit-refund systems, taxes,[1] subsidies and producer liability measures have expanded. Voluntary or soft instruments, including eco-labelling and voluntary agreements between industry and public administration, have also begun to be used more frequently.[2]

Interest groups and consumer behaviour

A growing public awareness of ecological issues has generated an increase in the number of citizen interest groups active in environmental protection. Today these groups account for more than four million members or about 5 per cent of the German population. They influence environmental policy making, both directly through their participation in public hearings on draft laws, and indirectly through information campaigns and publications.[3] As noted above, greater awareness of environmental issues has also been translated into consumption behaviour, with consumers increasingly focusing on the environmental and health properties of goods and services. However ecological enthusiasm does not always translate into behavioural changes. A survey conducted in 1992 showed that while many are concerned about the problems of air pollution, climate change and the destruction of the ozone layer, these worries only inspired 41 per cent of West Germans and 27 per cent of East Germans to pay higher gasoline prices in order to reduce the use of private transportation. Indeed if the price of gasoline were to double, as many as 74 per cent of West Germans and 66 per cent of East Germans stated that they would continue to drive to work in their own cars. On the basis of this survey, the environment minister concluded that while German citizens consider that environmental protection is important, as individuals they fail to take responsibility for pollution and perceive environmental protection to be the task of government and industry.[4]

Industry and commerce

The growth of environmental standards and regulations, consumer and political pressure, the need for internalising environmental costs and market opportunities are the principal factors that have motivated enterprises to adopt environmentally friendlier processes and to invest in the development of such products and processes. In addition, it has also been recognised that many savings can be made by using resources such as energy more efficiently.[5]

Industry and commerce have also recognised company environmental policies as a source of market opportunity and as an instrument for image building and advertisement. Coalitions between industry and ecologists, seemingly impossible little more than a decade ago, are also becoming increasingly common. A growing number of companies are introducing green product lines and publishing eco-balance sheets or green accounts covering the environmental aspects of their production processes and the life-cycle analyses of single products. Other companies have combined the themes of ecology and lifestyle. The Body Shop, a British-based cosmetics producer and retailer, for instance, employs a marketing strategy pledging the company's commitment to environmental and animal protection and to fair compensation for primary producers. The consumer thus not only buys a product, but an idea and a political identity. Fair trading associations are a further type of initiative. They highlight the objective of poverty alleviation in exporting economies as well as environmental friendliness and artisan goods or agricultural products, the most popular being the well-known 'fair-trade' coffees.

ENVIRONMENTAL LABELLING: THE BLUE ANGEL ECO-LABEL

The Environmental Label or Blue Angel label was introduced in 1977 and is awarded to products that are environmentally friendly relative to other products in their product category. It is a market-oriented instrument used on a voluntary basis and involves no binding requirements or bans.[6] The label is jointly sponsored by an independent jury made up of representatives from the scientific, business and environmental communities and consumer organisations. It also includes representatives of the German Institute for Quality Control and Labelling (RAL) and the Federal Environmental Agency (FEA).

The Blue Angel is awarded in two phases. During the first phase, product categories eligible for the award are designated. Subsequently the RAL organises expert hearings on selected product groups and the criteria for these groups, with the jury making the final decision. The Ministry for the Environment publishes the final decisions. In the second phase, manufacturers submit their applications for the label to the RAL, which undertakes an examination, on a case-by-case basis, of whether the producer fulfils the criteria. The FEA, the Ministry of the Environment in the producer's state, and on occasion the Federal Health Agency all participate in this process. The RAL and the producer then conclude a contract that entitles the latter to use the environmental label for a certain product.

In addition to these general criteria, product requirements are defined in detail for each of the product groups. Every three years these requirements are reexamined to determine whether the state of technology has developed and if the requirements need to be improved, whether other environmental aspects can be included and whether the standard of the environmental label has succeeded. When the standards initially set for the eco-label are used more frequently within a product group, the labelling for the group is suspended. This is done because the main objective of the system is to raise overall environmental quality through competition, and by the time the standards have generally been accepted this has been achieved. This also explains why product groups that are inherently environmentally friendly, such as bicycles, are not eligible for the label.

In order to use the label the manufacturer pays a flat fee of DM300 (US$180)[7] upon signing the contract with the RAL. An additional annual contribution, ranging from DM350 (US$210) to DM3980 (US$2388) is also levied, depending on the annual turnover of the company.

The label is well known among consumers. According to a 1987 survey, 79 per cent of all households knew of the label and 68 per cent understood its significance. In some product groups labelling has generated significant market changes. For instance the market share of low-pollutant coatings has increased from 1 per cent in 1981 to 20 per cent for commercial consumption and 40 per cent for household (do-it-yourself) consumption. While no systematic survey of these effects exists, it is evident that consumers increasingly consider environmental variables in their purchasing decisions, thus influencing product development.[8]

THE BLUE ANGEL AND FOREIGN PRODUCERS

The label is available to domestic and foreign producers alike, and all producers applying for the label have to furnish statements that their products comply with the criteria. For certain product groups, the RAL also requires neutral expert analysis. Upon using the label the producer agrees to observe the criteria and to reapply for the label if the product changes. Further control over the system comes not from product inspection by the RAL, but from the scrutiny undertaken by market competitiors and consumer associations. This type of informal market control has so far produced no evidence of non-compliance.

By mid-1992, 814 manufacturers had used the label for 3325 products in 75 categories, although no developing-country manufacturers were among them.[9] Considering the products currently labelled, the absence of manufacturers from developing countries is probably due to the fact that many of these products are not actually exported by developing countries. However, more worrying is the possibility that the gap is due to a lack of information, particularly in countries outside Western Europe, about the market advantages that can be generated by the label and the fact that foreign manufacturers are eligible to use the label. Apart from these difficulties, it also appears that serious trade obstacles are being created for foreign exporters as the scheme is designed mainly for domestically produced goods and services. For instance the Brazilian pulp and paper industry encountered problems with the German Blue Angel because it favoured a percentage of recycled paper in labelled products, but failed to take into account whether this type of recycled paper was available to overseas suppliers. Neither did the scheme take into account the benefits to the Brazilian environment of paper manufactured from sustainably managed forests.[10]

It seems clear that environmental product requirements are likely to grow in future and eco-labelling will continue to change production and marketing practices in OECD markets. Consequently manufacturers in developing countries will need to study the possibility of using such labels. This may require adaptation of their products to the criteria set in OECD countries, or persuading labelling agencies to adjust the criteria if they are inappropriate for developing countries. This also raises important questions regarding the costs of adaptation and access to environmentally friendly production technology. However, in the long run, participation by firms from developing countries in the Blue Angel will improve market opportunities and earnings from the German market and will facilitate access to other markets, especially in OECD countries.

Information about these labelling schemes is frequently published and is made available through international chambers of commerce. Assistance and cooperation may also be available in this regard for developing countries through the German Society for Technical Cooperation (GTZ).[11]

THE TEXTILE INDUSTRY'S ECO-LABELLING PROJECTS

Unlike the Blue Angel Label, textile eco-labelling was not initiated by the government but by industry itself. In June 1992 the Association of Textile Producers (Gesamttextil e. V.) founded the 'Verein für verbraucher- und umweltfreundliche Textilien'.[12] The task of this Association is to 'promote the consumer-friendliness of textile products and foster production processes which do not jeopardise the environment'.[13] Two types of label have been proposed for this purpose:

- the MST ('Marke schadstoffgeprüfter Textilien') product label for textiles that reach the final consumer, and
- the MUT ('Marke umweltschonender Textilien') product label for intermediate textile products that are manufactured in an environmentally benign way and do not enter the retail market.

The criteria for these textile labels is based on the eco-label award scheme of the European Union, although the Association of Textile Producers would have preferred more stringent criteria in the interests of protecting the health of consumers.[14] The MST label can be awarded for all textile and clothing products with the exception of textile flooring materials. Textiles labelled with the MST must not contain certain carcinogenic chemicals or be produced using chlorinated carriers during the dying process. They must also not exceed quantitative ceilings for heavy metals (contained in zippers, buttons and some dyes), formaldehyde, pesticide residues, or the residues of chemical fertilisers and defoliants. In addition they must not give off dye in contact with saliva and sweat if produced for children up to two years of age, or contain a pH-value different from that of human skin.[15]

Since the MST label was introduced in 1993, 66 manufacturers have submitted 200 applications to use the label, and about 60 per cent of applicants have so far been awarded the use of the label.

The draft award criteria for the production-related MUT label relate to fibre production, textile processing and disposal. For fibre production, the criteria include norms for the use of chemicals, such as ceilings on the use of chemical fertilisers and defoliants in cotton production, and norms for the use of pesticides in wool production, as well as for the treatment of effluents from wool washing and scouring. For textile processing, there are limitations on the harmful dyes and chemicals that can be used, norms for the biological treatment of industrial sewage at the production site, regulations on air pollution modelled after German technical instructions and the control of noise. Criteria regarding disposal of wastes, ease of product recycling, landfill disposal, incineration and the presence of heavy metals in buttons, zippers and metallic compound dyes are also regulated.[16]

A number of problems have been encountered with regard to the MUT. This label was created to cover the full life-cycle of intermediate products, including fibre production, use and disposal. However, since the variables to be considered are diverse, the Association has faced difficulties in devising reliable monitoring procedures. While the disposal qualities of a textile product are relatively easy to assess, verifying the application of production norms for the primary inputs of wool and cotton poses serious problems, especially if factors such as the treatment of effluents from wool washing and scouring at the site of production are to be included. Earlier attempts to include textiles under the Blue Angel scheme were abandoned because of these difficulties. Moreover the Danish working group that is currently determining criteria for textile products (T-shirts and bed linen) as part of the European Union's eco-labelling scheme has encountered similar difficulties. For this reason the Association does not expect to be able to use the MUT in the near future.

TEXTILE ECO-LABELLING AND FOREIGN PRODUCERS

The Association proposing textile eco-labels has restricted its membership to textile producers located in the countries of the European Union and EFTA. The MUT production and process-related label, in its current draft form, is also intended to be reserved exclusively for producers in these countries because it is only in these countries that the Association can be assured of the quality of the production process.

While critics of these labels see this exclusivity as a trade barrier, the Association has emphasised that non-members are eligible for a label as long as their products comply with the label criteria. However compliance is undertaken by textile institutes at a cost of DM2000 (about US$1200), which has to be covered by the producer. Presently 10 per cent of all labels have been awarded to producers in France, Italy, Austria, Switzerland and Greece. According to the German Textile Producers association, no interest in the MST has so far been expressed by textile manufacturers from developing countries.[17] The German Association is also currently looking into the possibilities of standardising the MST label with other European textile labels.

Gesamttextil's labelling plans have encountered strong opposition, in particular from retail traders and the clothing industry, which procure a large share of their supplies from East Asian countries. It has been asserted that introduction of the label will have a profound impact on international trade, affecting some 70 000 manufacturers and exporters in East Asian countries alone.[18]

The major objection to the establishment of the MST and MUT eco-labels is that their application will violate Article IX of the GATT, because the awarding of criteria for these labels will constitute a discrimination against manufacturers and exporters in third countries. Moreover the labels appear to constitute non-tariff barriers to trade and are thus inconsistent with Article XI of the GATT and Article 30 of the EEC Treaty. In addition, it is argued that the awarding criteria do not, or do not exclusively relate to the product as such, but to process and production methods, which again is contrary to international trade rules.[19]

The Association has however argued that its labels constitute private rather than government measures and therefore should not be considered trade barriers under WTO rules. This argument has nonetheless been challenged because, although these eco-labels are a private initiative, the Federal Environmental Agency created a *de facto* government-initiated measure by inviting industry to initiate the labelling programme.[20]

In fact the MUT, as it is currently conceived, represents at least an informal barrier to trade, since countries outside the EU and EFTA are by definition excluded from the label. While this may not violate WTO rules, it could constitute a violation of anti-trust laws.

Furthermore, although the MST admits producers from outside Europe, its norms may be difficult and costly for exporters from developing countries to comply with. The industry argues that eco-labelling textiles is an advertising instrument that helps compensate for the comparative advantage that industries in developing countries have by operating with lower labour costs and lower environmental process and product standards. Due to competition the textile industry in Germany has shrunk by about 50 per cent over the past decade, while at the same time enterprises have been required to invest large sums in environmental protection. The industry now hopes to capitalise on these adjustments by promoting textile sales through eco-labelling. The industry argues, moreover, that given the growing market preference for goods produced in an environmentally friendly manner, textile labelling meets the consumers' need for information. Textile producers in developing countries tend to interpret the growing use of this type of labelling in Germany as an attempt by German industry to protect its markets and amortise investment.

The success of the MST and other textile labels will largely depend on their acceptance by consumers and the retail trade. The market, however, is not homogeneous and price remains the chief determinant of overall demand. Other variables determining the size of the market for eco-textiles are short-term economic expectations. In the long run, environmentally friendlier and healthier products will be favoured by those with a 'green' product preference, while goods without eco-labels may be disadvantaged. Exporters to Germany and to other OECD countries should closely watch these trends and make reasonable adjustments in order to maintain or gain competitiveness in higher-quality market segments.

ECO-PACKAGING

The objectives of the German packaging ordinance of 1986, officially titled the Waste Avoidance and Management Act, include the prevention and reduction of waste and the improvement of waste recovery, reuse and recycling, so as to reduce stress on landfill capacities. Packaging waste currently represents about half of the volume and one quarter of the weight of municipal waste.[21] The ordinance is based on the polluter-pays principle and gives preference to the objective of avoiding waste over reuse and recycling. It does not permit the incineration or landfill disposal of packaging waste if reuse or recycling are technically and economically feasible and if markets exist or can be built for secondary raw material. Only residual waste that cannot be reused or recycled is permitted to be disposed of.

The ordinance requires that producers and retailers take back packaging waste for recovery and disposal, and that polluters bear the costs of waste management. This provides an incentive for industry to reduce the volume of material produced and also shifts the responsibilities of waste management onto the private sector. The requirements of the ordinance came into effect on 1 December 1991 for transport packaging, on 1 April 1992 for secondary packaging, and on 1 January 1993 for sales packaging.

The ordinance does not specify what materials should be used in packaging, and it allows service enterprises, rather than the producers themselves, to collect and recycle the packaging material. This means that producers exporting to Germany are not required to collect their packaging or take it back home, but can instead join an organisation that will collect and recycle the packaging waste on their behalf. A private company, the Dual System (DSD), has been set up to collect and recycle sales packaging, while other organisations collect and recycle transport packaging waste.

In practice the recyclability of packaging material is limited by the existing capabilities of the recycling industry, rather than by the properties of the material actually used. Since recycling technologies are being adapted to new needs and to growing capacity requirements, material specifications are constantly evolving. For products originating outside Germany, importers normally verify with recycling companies that the packaging material they are using can be recycled. The recycling companies also inform producers about their material requirements and issue the necessary documents to confirm that the packaging material can be recycled and complies with the legal requirements.

Sales packaging, the Green Dot and the Dual System

The Dual System or DSD, is a group of about 400 enterprises that services the entire country and operates parallel to the public waste collection service. The DSD collects sales packaging waste directly from households if it is marked with its license stamp: the 'Green Dot'.[22] Glass, tinplate, laminated cardboard, paper and paperboard are all guaranteed by recycling contractors and are generally granted a green dot. However the capacity to recycle all types of plastic, laminated plastic and aluminium beverage packaging (with the exception of aluminium beverage cans) does not yet exist.

The Dual System is financed through licensing fees paid by member producers, but in practice consumers have financed the activities of the DSD as well.[23]

Since its introduction, the Dual System has been extremely controversial and the government has on various occasions threatened to withdraw its license. The capacity to recycle and financial problems have plagued the system. In the past, guarantees have been issued even though recycling capacities were insufficient. Germans are fervent collectors of used packaging and the system has to some extent been a victim of its own success. Packaging waste collected in recent years has been four times the volume expected.

Due to insufficient capacity, the recycling operation set up by the plastics industry has collapsed amid accusations that it has been exporting waste rather than recycling it. To remedy this, collection has been permitted outside the DSD. In addition, some recycling quotas have been lowered in response to earlier unrealistic targets.[24]

Growing waste exports are another outcome of the shortage of recycling capacity. Neighbouring countries have complained that German waste exports are flooding their own infant recycling industries for plastics, paper and board. In France, calls have been made for the introduction of import authorisations and a ban on waste imports.[25] Import bans for plastic waste have been imposed by the governments of Taiwan and Indonesia. In addition, a significant proportion of European plastic refuse, including waste from DSD partners, appears to have been sold as secondary raw material to firms in Malaysia, the Philippines and Indonesia where, according to reports, it is being recycled into sandals and motorcycle helmets for the Asian market. Moreover allegations have been made that some DSD subcontractors charge fees for recycling, only to export the Green Dot packaging waste and dispose of it in landfills. According to one of the largest European waste brokers, 15–25 per cent of plastic waste is not being recycled and instead finishes up being dumped in developing countries.[26] Germany is considered the world's leading exporter of waste.[27]

The costs of operating the DSD, which still faces a financial crisis, have also expanded as a result of increases in prices demanded by waste management companies and poor payment of fees by member producers.

Furthermore environmental groups have criticised the Dual System's exclusive focus on recycling rather than reducing waste, which they claim is undermining the true intent of the ordinance. The production of waste will not diminish in a situation where the recycling industry benefits from growing rather than diminishing quantities of packaging. Moreover interest groups have demanded a rethinking of the strategy of 'recycling at all cost' and have sought the adoption of incentives favouring products without packaging and the more environmentally friendly 'deposit-refund' systems.[28] To support such schemes, it has been suggested that a progressive tax on packaging material be introduced. Critics of deposit-refund systems, however, assert that such systems can constitute barriers to trade, especially if they are made compulsory.[29] Environmentalists have also criticised the success of the Green Dot as a marketing instrument, claiming that the dot implies the packaging is environmentally friendly, when in fact it is not.

On the positive side, however, the volume of packaging waste does appear to have declined slightly in recent months and the growing costs of recycling have actually led to developments towards smaller packages and the use of fewer materials in packaging.

The trade impact of measures for recycling transport and sales packaging

The packaging ordinance affects all exporters to Germany. However transport packaging is the most critical type of packaging for developing countries, since most of their products are packaged for sale in the importing country only. Important issues arising from measures related to sales packaging are the importation of waste and the effects of production based on secondary raw materials that takes place in some developing countries.

Since the ordinance came into effect, exporters have experienced a number of problems regarding transport packaging. A major difficulty is the lack of information about the new requirements, especially new recycling specifications and the capacity of certain recycling industries to process the waste. It seems clear that improvements must be made to the information channels between the recycling industry and producers.

Producers in developing countries are also concerned about the failure of recycling industries to take into account the characteristics of their products. Recycling requirements have predominantly been based on domestic environmental resource endowments and constraints in the country applying the packaging requirements. For instance, Colombian flower exporters have experienced considerable delays because of recycling difficulties resulting from their use of wooden pallets, metal staples and laminated materials. These problems have impacted negatively on their export competitiveness.

The case of jute

Exporters have also experienced a number of difficulties with jute packaging. This has led to growing pressures to substitute other materials for jute, such as polypropylene plastic. However, should the use of plastics prevail over jute, this shift would not only entail the substitution of a material that is relatively environmentally-friendly with a more damaging one, but it would also deprive producer countries such as Bangladesh of their earnings from jute exports.

On the German side of this issue, several factors are reported to have contributed to the jute problem. First, jute sacks are used extensively in the transportation of certain foodstuffs, especially coffee, cacao, legumes, spices and dried fruit, as well as tobacco, wool and cotton. As residues of pesticides are found in much of the jute packaging, worries have been voiced about contamination of these foodstuffs. However this worry may hold less for coffee, which is typically exported as beans and is thus protected by its thick shell. Secondly, residues of batching oils utilised to render jute fabric more supple and prevent it from rotting are also difficult to extract and prevent composting and recycling of jute into filling and other materials. Consequently, because the main disposal option for jute is to incinerate it, this makes other materials, even plastics, appear preferable. In addition there appear to have been problems in finding recycling firms willing to handle jute because of the low quantities involved and the poor economies of scale.[30]

These difficulties are gradually being alleviated, especially since recycling capacities for jute are easier to create than for materials such as plastics, and there is the possibility of burning jute for energy recovery with new and environmentally friendlier incinerating technology. The substitution of batching oils by less harmful substances is another option. Moreover jute producers may consider applying for the Blue Angel label, which may raise the price for jute, reduce recycling costs, and provide jute with greater demand on the secondary market. It should also be noted, however, that independently of the material-related problems, container shipping is currently gaining ground over transport in jute sacks, particularly for coffee.

OPTIONS FOR EXPORTERS WITH SPECIAL REFERENCE TO SOUTH ASIAN COUNTRIES

For exporters to the German market, new environmental measures raise important questions about their impact on exports and the need to adapt products to new and evolving specifications. A chief concern is the difficulty of obtaining the necessary information about environmental requirements and their lack of transparency, which can form significant barriers to trade. It is a sobering thought to consider that at a time when tariff barriers are being lowered on a global scale, developed countries appear to be engaging in an insidious 'green trade war' to protect their domestic industries.[31]

The product groups that are most affected, or could be seriously affected in the future by German environmental and health-related requirements include fishery products, other food and beverage products, coffee and tea, wood and wood products, paper, fibres, yarn, textiles, garments, and especially cotton products, leather products and packaging. For South Asian countries, the volume of export earnings that could be affected by the environmental measures in these product groups amount to US$1807 million or 61.8 per cent of the total export earnings from the German market in 1992.

Garments comprise 67.5 per cent of earnings within the group of potentially affected exports from these countries and thus make up by far the largest product group in this category. Export earnings in this group amounted to US$1219 million in 1992, or 41.7 per cent of the total export earnings of South Asian countries. If cotton, cotton fabrics and other yarn, fibres, fabrics and textile articles are added to this figure, it would comprise 85 per cent of all potentially affected products, and its share of total export earnings would jump to 52.5 per cent, or US$1537 million.

The chief concern of South Asian producers exporting to Germany must be the environmental labelling measures being implemented by the textile industry. The degree to which German consumers shift their demand in response to these measures may affect about half of all textile exports to Germany. Should textile labelling have a significant trade effect, the gains expected from the phasing out of the Multifibre Arrangement may, at least in part, be wiped out. South Asian exporters need to focus their attention on developments in this sector.

It is vital for the needs of all developing countries that efforts be made to increase the transparency of these new environmental measures through official and private channels. Governments need to aim for mutual recognition and harmonisation of these measures, preferably through the International Standardisation Organisation (ISO), while the WTO must tackle questions of 'green' protectionism. It should be noted, however, that although mutual efforts to recognise standards would be effective in the medium term, producers in developing countries tend to define their strategies and responses in the short term.

CONCLUSIONS

Traditionally, developing countries have been less affected by product standards and norms since they have been applied primarily to high-tech and leading edge technology products, an area in which developing countries were less competitive. Today, however, environmental and health characteristics are of growing importance and affect the traditional export sectors as well as the marketability of products and services. Increasingly, these properties are becoming important determinants of competitiveness and pose a challenge to the innovative capabilities of manufacturers in developed and developing countries. Export-oriented industries have no alternative but to adjust to new government regulations and buyers' specifications, even if their scientific justification is frequently questioned.[32] As numerous standards still differ between markets, exporters should aim to adapt their products to the highest existing standards. This strategy will be costly, but it is the best way to build up competitiveness and will, in the long term, ensure both greater flexibility in a large number of markets and higher export earnings.

Exporters in developing countries must keep up with ongoing environmental discussions and seek access to data bases documenting environmental measures in importing countries so that they can prepare for shifts in regulations and demand. Moreover governments of exporting countries need to participate actively in the ongoing efforts to mutually recognise and harmonise environmental standards at the International Standardisation Organisation. Participation at this level offers opportunities to urge greater transparency. UNCTAD too offers information about the impact of environmental measures on developing countries and their trade.

Producers in developing countries must explore the possibilities of eco-labelling and other environmental initiatives. There is real potential for a jute eco-label as well as plenty of opportunities for linking in with fair trade and environmentally orientated enterprises in developed countries. Efforts must be expanded to find mutual recognition of eco-labels between developed and developing countries.

In order to advertise the inherent environmental qualities of products, especially traditional export products, and to capitalise on improvements in the environmental quality of their products, as well as to create new trading opportunities, manufacturers in developing countries could conceptualise new labels geared towards export markets. Labels can be introduced for very specific export products, as demonstrated by Indonesia with its environmental label for rattan furniture.

New labels could also include criteria other than just environmental ones. For instance a 'fair trade label' indicating to the consumer the developmental impact of a product appears to be a worthy initiative. The experience of existing fair trade labels, such as the Max Havelaar coffee label, should be analysed and adapted to the needs of producers in other sectors. Cooperation with institutions such as the German GEPA,[33] which applies transparency, fair trade and environmental criteria to its marketing, could be a useful first step. Surveys of other existing labels and the marketing channels for environmentally sound products in major export markets would also provide useful assistance for exporters in developing countries.

Producers in developing countries also need to investigate the possibilities for technical cooperation with technical institutes in adapting their products and marketing strategies to the new environmental standards and the criteria set by existing eco-labels. In Germany, cooperation could be solicited through the German Society for Technical Cooperation (GTZ) and the Federal Environmental Agency (FEA).

Finally, it is up to firms in developing countries and their governments. They may see the rise of environmental regulations and standards in developed countries as 'green' protectionism – and indeed in some cases they may be right. But the environmental crisis in OECD countries is not just a perception. It is a real threat and there will continue to be enormous pressure from consumers and taxpayers for industries and governments to march to the environmental tune. For whatever purposes environmental initiatives are taken in these countries, they are bound to have some adverse effect on commerce and trade. Consequently it is up to developing countries and their firms to avoid the negative effects, make complaints where there are real cases of protectionism and take advantage of whatever opportunities present themselves.

Notes and references

1. See Federal Ministry for the Environment-FME (ed.) *Environmental Protection in Germany*, (Bonn: FME, 1992) (National Report of the Federal Republic of Germany for the United Nations Conference on Environment and Development in June 1992 in Brazil), p. 91; Imme Scholz and Jürgen Wiemann, *Ecological Requirements to be Satisfied by Consumer Goods – a New Challenge for Developing Countries' Exports to Germany* (Berlin: German Development Institute, 1993), pp. 1–2; UNCTAD, *UNCTAD's Contribution, within its Mandate, to Sustainable Development: Trade and Environment – Trends in the field of trade and environment in the framework of international cooperation* (Geneva: UNCTAD 1993, (TD/B/40(1)/6), p. 9; OECD, *Environmental Performance Reviews – Germany* (Paris: OECD, 1993), pp. 109–15.
2. See OECD, *Environmental Labelling in OECD Countries* (Paris: OECD, 1991), p. 12, and Laurence Schlosser, René Vossenaar and Simonetta Zarrilli, 'Environmental Regulations and Trade' (Geneva: UNCTAD, October 1992 (internal paper)), pp. 9–10.
3. OECD, *Environmental Performance Reviews – Germany* (Paris: OECD, 1993), p. 20.
4. Bundesumweltminister, 'Repräsentative Bevölkerungsumfrage zu Einstellungen zum Umweltschutz 1992 in Ost- und Westdeutschland' Bonn: Bundesumweltministerium [Federal Ministry for the Environment] (Press release 107/92; 30 December 1992).
5. See UNIDO, *Proceedings of the conference on ecologically sustainable industrial development – Copenhagen, Denmark, 14–18 October 1991* (Vienna: UNIDO, 1991), pp. 54–5.

6. See RAL Deutsches Institut für Gütesicherung und Kennzeichnung e. V., *The Environmental Label Introduces Itself* (Bonn: RAL, August 1992), pp. 4–8.
7. Calculated at the exchange rate of 30 March 1994, as quoted in the *Financial Times*.
8. See Umweltbundesamt (FEA), *Das Umweltzeichen stellt sich vor* (Berlin: Umweltbundesamt, 1990), pp. 25–9.
9. Among the manufacturers who have signed contracts to use the label, 15 per cent (121 manufacturers) are foreign companies of the following origin: Austria (13), Belgium (5), Denmark (6), Finland (1), France (22), Great Britain (11), Italy (16), Liechtenstein (4), Netherlands (21), Norway (1), Spain (1), Switzerland (15), Yugoslavia (2). See Federal Environmental Agency, 'Information Sheet on the Environmental Label' (Berlin: Federal Environmental Agency, 1992), p. 4.
10. See GATT, *Trade and Environment Bulletin*, no. 4 (Geneva: GATT, 26 November 1993).
11. German Society for Technical Cooperation (Gesellschaft für Technische Zusammenarbeit)
12. Association for the Promotion of Textiles Friendly to Consumers and the Environment, referred to hereafter as the Association.
13. See 'Satzung des "Verein für verbraucher- und umweltfreundliche Textilien e. V"' (Frankfurt: Verein für verbraucher-und umweltfreundliche Textilien e. V. (1992)) (articles of the Association for the Promotion of Textiles Friendly to Consumers and the Environment; manuscript).
14. See 'Council Regulation (EEC) No 880/92 of 23 March 1992 on a Community eco-labelling award scheme', in *Official Journal of the European Communities*, no. L 99 (11 April 1992), pp. 1–7.
15. See International Textiles and Clothing Bureau, 'Environment Protection and Textiles' (Geneva: ITCB, 22 June 1992, information sheet).
16. Ibid.
17. Author's interview with Mr Virnich, a spokesperson for Gesamttextil (14 February 1994).
18. See Foreign Trade Association-FTA, 'Gesamttextil's Eco-Labels in Violation of International Law Norms' (Brussels: FTA, 21 July 1992, Memorandum).
19. See Gilbert Gornig, *The Consistency of Eco-Labels with International Law, especially with the GATT* (Cologne: Aussenhandelsvereinigung des Deutschen Einzelhandels e. V., 1992), pp. 16–24.
20. Ibid., p. 19.
21. See OECD, *Environmental Performance Reviews – Germany*, p. 58.
22. The Green Dot serves as a license mark and a signal to consumers indicating that the marked packaging should be thrown into the yellow garbage container distributed to households by the Dual System rather than into the communal cans. Thus the Green Dot is purely a financing instrument for the disposal of non-reusable packaging rather than an eco-label identifying an environmentally friendly product, such as the Blue Angel.
23. See *Der Spiegel*, 'Recycling ist nur der zweitbeste Weg', vol. 47, no. 25 (21 June 1993), pp. 34–50.

24. *Süddeutsche Zeitung*, 'Töpfer beschränkt das Duale System', 29 December 1994, p. 17.
25. See David Gardner, 'France May Ban German Waste Imports', *Financial Times*, 30 June 1993, p. 16.
26. See *Der Spiegel*, 'Recycling ist nur der zweitbeste Weg', vol. 47, no. 25 (21 June 1993), pp. 34–41.
27. See Martin Runge, *Milliardengeschäft Müll. Vom Grünen Punkt bis zur Müllschieberei – Argumente und Strategien für eine andere Abfallpolitik* (München: Pieper, 1994), p. 169.
28. See *Der Spiegel*, 'Recycling ist nur der zweitbeste Weg', p. 34.
29. See John Hunt, 'World Business Community Fears 'Green' Dumping', *Financial Times*, 12 April 1991, p. 5.
30. According to Delbrück, recovery has reached 50 per cent, the same capacity as for plastic packaging. Kilian Delbrück, *Eco-Packaging, 'Green Dot' and 'Blue Angel': The German Case*, paper presented at the Workshop on Trade and Environment, Bogotá, 19–21 October 1993.
31. See Rajiv Shirali, '"Green" Signal to Restrict Trade Traffic', *The Economic Times New Delhi*, 21 February 1993, p. 7.
32. See Jürgen Wiemann, 'Green Protectionism – A Threat to Third World Exports? The case of Indian leather and textile exports facing tighter environmental standards on the German/European markets' (Berlin: German Development Institute, 1993), pp. 6–14.
33. German Society for the Promotion of Partnership with the Third World Ltd.

5 Environmental Standards: Relocation of Production to the SAARC Region

Roland Mollerus*

The linkages between environmental standards and international competitiveness have in recent years received widespread interest.[1] It is commonly believed that where environmentally orientated product standards have an impact on competitiveness, this will normally be to the benefit of the importing country. The impact of environmentally orientated process standards, however, is less clear. Conventional wisdom indicates that the need to comply with more stringent domestic environmental process standards will adversely affect domestic industry vis-à-vis foreign competitors, because of the impact of compliance on the costs of production. It is feared that this is leading to the relocation of industries from countries with high environmental process standards to countries with lower ones.[2]

This compliance-cost argument indicates that a competitive advantage accrues to industries that operate in countries with lower environmentally orientated process standards. This situation is reinforced by the fact that while it is mandatory for domestic firms to comply with national process standards, these cannot at present be applied to imported products through border trade measures. On the other hand, however, there are disadvantages to firms in countries with lower environmental process standards where they do not have the technological capability to meet the higher voluntary standards or stringent consumer tastes being formed in foreign markets.

Generally speaking, the costs of complying with environmentally orientated process standards, as well as the costs and benefits of influencing voluntary standard setting, are unevenly distributed across industries. Clearly there are winners and losers, depending on the specific characteristics of a particular industry. Nonetheless, whatever the impact of environmental standards on competitiveness and trade performance, some sectors will be more sensitive to environmental factors than others.

This chapter first quantifies the significance of these so called 'polluting industries' for the exports of the South Asian Agreement on Regional Cooperation (SAARC) countries. It does this by analysing the export shares of trade in products from manufacturing industries that would have to incur relatively high adjustment costs if environmental standards were raised.[3] The chapter then examines whether 'ex-post' indicators of international competitiveness demonstrate an increased or decreased competitiveness for polluting industries over time.

TRENDS IN THE RELATIVE IMPORTANCE OF POLLUTING INDUSTRIES FOR SAARC ECONOMIES

The identification of polluting industries in SAARC economies has been based on data for the pollution abatement and control expenditures of United States manufacturing industries, as reported in Manufacturers' Pollution Abatement Costs and Expenditures. This approach was taken because systematic data concerning the pollution control costs in SAARC countries is not available from other sources.[4] Moreover, for the purposes of this study only those manufacturing industries that have the highest levels of pollution abatement and control expenditure were considered as polluting industries.[5]

Export data was examined to identify whether there have been significant changes in world trading patterns of products manufactured by polluting industries. A comparison was then made of the trade in products from polluting industries based in North America, the European Union (EU), Japan and SAARC countries. In order to compare the performance of SAARC countries with other developing countries, the exports of products from polluting industries based in a group of Latin American countries as well as in a group comprising all developing countries were also examined. The inferences drawn from this data are outlined below.

The relative importance of exports of products from polluting industries

Between 1982 and 1992 the overall share of world trade (by value) held by products from polluting industries actually decreased from 16 per cent to about 13 per cent. In contrast, in SAARC countries the percentage of products from polluting industries compared with total exports grew slightly during this period. In developed countries, on the other hand, the share of these exports dropped in line with global trends. It appears then that the significance of products manufactured by polluting industries in the EU, Japan and North America have diminished while these types of export have been gaining in importance for developing countries. This is especially the case for Latin American countries, where the export of products from polluting industries comprise 34 per cent of total exports.

Growth rates in the value of exports from polluting industries, compared with total manufacturing exports, indicate that between 1982 and 1992 the rates of growth in polluting industry exports for developing countries were higher than the growth in manufacturing exports as a whole. On the contrary, the growth rates of developed countries show an opposite trend. There has also been a relatively high growth rate (22 per cent) in the value of exports from polluting industries based in SAARC countries when compared with other developing countries.

The export share of products from polluting industries based in SAARC countries, compared with total manufactured exports to the EU, Japan and North America, has also increased in importance over the past 10 years, especially in the percentage of exports to the EU.

Although SAARC countries are exporting fewer products from polluting industries in absolute as well as relative terms when compared with other countries, there appears to be a trend towards growing importance of these types of export for the SAARC region. This does not imply that SAARC countries are capitalising on low environmental standards. Instead it seems that they are, as part of the industrialisation process, progressively specialising in process industries (such as chemical and metal industries), which are more likely to experience higher pollution abatement costs than discrete part manufacturers (such as electronics).[6]

These findings do not support a conclusion that the increasing share of exports from polluting industries is due to lower environmental standards. Neither do they prove that polluting industries have migrated from countries with high standards to developing countries because of differences in regulatory systems. Moreover, changes in prices may actually alter the conclusions provided above. For example the prices of agricultural products may well rise as a result of the Uruguay Round whereas those of polluting industries may fall. This may change the above trends. It should also be noted that changes in the relative prices of products from polluting industries vis-à-vis non-polluting industries could well be responsible for the changes in the relative shares of these industries, without reflecting any real changes in industrial activity.

Indicators of competitiveness in polluting industries

In seeking to determine the competitiveness of polluting industries in SAARC countries, an analysis was undertaken of the exports of products from these polluting industries. Attention was focused on changes in market structure over the period 1982 to 1992.[7] The approach taken was based on a classification of export products into four clusters: two clusters indicating whether OECD trade in a specific product is increasing or decreasing relative to the OECD's total imports (plotted from the horizontal axis of Figure 5.1); and two clusters indicating whether or not the share SAARC countries have in OECD imports of these products is increasing (plotted from the vertical axis of Figure 5.1).[8] On this basis, the following classification of export products from polluting industries was made:

1. 'Retreats' represent declining products in world trade in which the export products of SAARC polluting industries are losing their market share.
2. 'Waning stars' represent declining products in world trade in which the export products of SAARC polluting industries are gaining in market share.
3. 'Missed opportunities' represent increasing products in world trade in which the export products of SAARC polluting industries are losing their market share.
4. 'Rising stars' represent increasing products in world trade in which the export products of SAARC polluting industries are gaining in market share.

By attributing the export products of polluting industries a product-specific combination of competitive position and market attractiveness (based on market share growth rates over the period 1982 to 1992) each product can then be classified into one of these four categories.[9]

However, since competitiveness is determined by the relationship between a number of factors that cannot be easily quantified, it should be noted that while this type of analysis reveals competitiveness, it does not provide an explanation for it.[10] Nevertheless the matrix offers a simplified view of the competitiveness of SAARC countries.

The results of this analysis suggest that half the products from polluting industries exported from SAARC countries are grouped under 'waning stars' and less than one third are competitive in a growing export market. Less than one third are uncompetitive. Eight of these industries are combining their unfavourable competitive position with a decreasing market share in world trade (retreats), while exports from 20 industries are characterised by a low degree of competitiveness in a world market of increasing demand (missed opportunities). Nevertheless the share SAARC countries have of total OECD imports of these types of product has actually been increasing for more than 95 per cent of trade in these products (Figure 5.2).

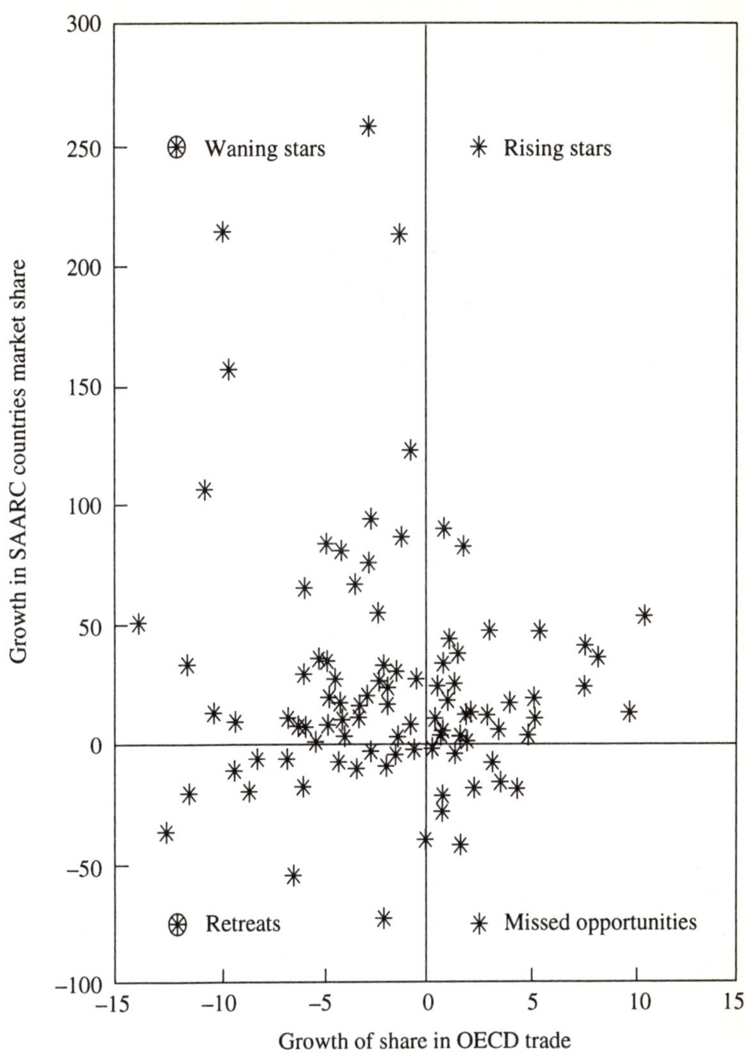

Source: Roland Mollerus, based on the COMTRADE database.

Figure 5.1 Competitiveness matrix of SAARC countries' polluting industries, 1982–92

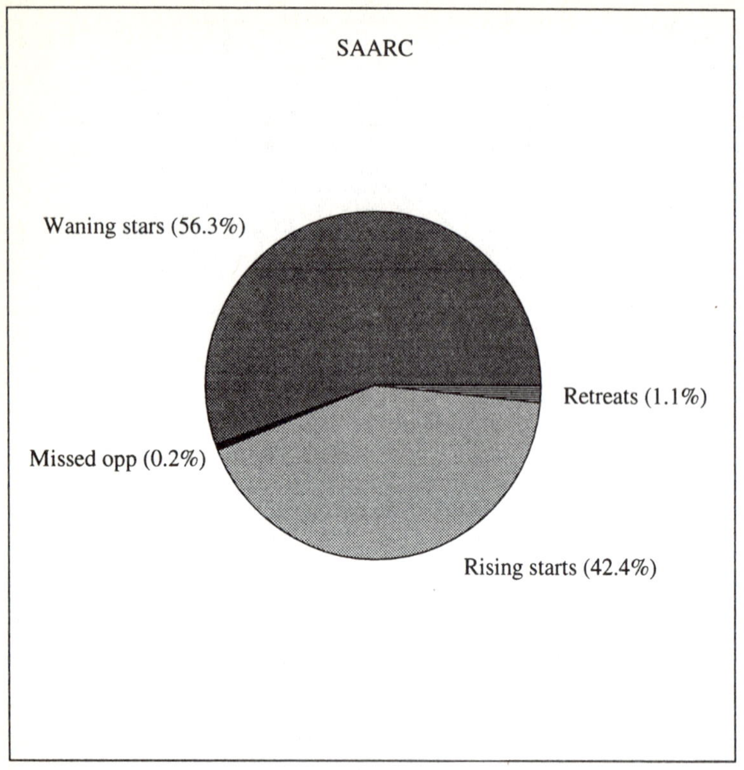

Source: Roland Mollerus, based on the COMTRADE Database.

Figure 5.2 Relative importance of SAARC exports of polluting industries to OECD markets, based on the competitiveness classification of Figure 5.1 (percentage of total SAARC exports from polluting industries to OECD markets, 1992)

Implications for the SAARC region

Overall, it is difficult to conclude whether differences in environmental standards have affected competitiveness in polluting industries and induced relocation of these industries to the SAARC region. Although lax environmental standards may have contributed to an increase in competitiveness in specific instances, empirical evidence rarely indicates that differences in environmental standards are a major source of competitive advantage and/or industrial relocation.

Despite the lack of convincing evidence of the relocation of certain industries for environmental reasons, trade and competitiveness indicators show an increase in the importance of polluting industry exports for the SAARC region. Consequently, if these exports remain important for the SAARC region, any capital costs imposed by changes required to comply with new national or international standards will be high.

The cleaner technologies required to comply with higher environmental standards may not be readily available in the SAARC countries, or installing them may necessitate the scrapping of existing facilities. Moreover SAARC countries may face uncertainties about what cleaner technologies they should invest in. Cleaner technologies are also to a certain extent determined by the regulatory regimes of OECD countries. Consequently progress in developing and implementing cleaner technologies may be extremely rapid, depending on the changes being made in OECD regulatory regimes. It may be difficult for SAARC countries to keep up.[11]

The transfer of hazardous or obsolete technologies from developed countries to the SAARC region may also intensify environmental problems in the SAARC region. Access to information about the hazards of imported technologies and their cleaner alternatives would assist operators to improve their environmental performance. In this context, it is important for the World Trade Organisation Committee on Trade and Environment to consider the trade in hazardous technologies as an item of its deliberations on 'Trade in domestically prohibited goods'.

If implemented, new environmental standards for polluting industries in the SAARC region should be kept to a minimum level and only raised progressively so as to avoid losses in comparative advantage. If standards are set too high or introduced too abruptly, they will tend to reduce the competitiveness of existing industries, thereby inhibiting the ability of these countries to mobilise, through international trade, the resources needed to finance the investment required for sustainable development and economic growth. It should be kept in mind that only through economic growth will there eventually be a shift to cleaner production.

CONCLUSIONS

There is a persistent argument in the literature on trade and the environment that posits that diverging environmental standards provide a competitive advantage to polluting industries in certain developing countries. With regard to SAARC countries, however, a statistical analysis of exports from their polluting industries indicates that polluting industries account for only a very small, although increasing, share of SAARC countries' manufactured exports. Evidence from competitiveness indicators, based on market shares, show that these polluting industries have, in general, increased their comparative advantage. There is however no clear evidence of relocation of polluting industries to the SAARC region due to lower environmental standards.

Although environmental reasons for relocation and competitiveness in polluting industries have played only a modest role to date, it should be borne in mind that environmental costs look like becoming more significant in future as standards become more strict. For example carbon taxes may have a significant impact on international competitiveness. New US regulations such as the Clean Air Act Amendment of 1990 and prospective new regulations in other developed countries, such as those dealing with greenhouse gas emissions, may involve larger costs and alter the conclusions of research undertaken in the past.

Nevertheless the efforts of SAARC countries to internalise the environmental costs of their polluting industries deserves more positive international support. International cooperation should aim at accelerating development, maintaining an open trading system, providing the mechanisms for 'clean' technology transfer as well as building an institutional capacity to integrate trade and environmental policies in the framework of national policies for sustainable development. It would be a negative step to allow claims of competitive disadvantage to be used to raise barriers to international trade.

Notes and references

* The views expressed in this chapter are those of the author and do not necessarily reflect those of the United Nations Conference on Trade and Development.

1. See also Chapters 2 and 3 of this book.
2. The relocation of industries may be defined as: (i) firms in developed countries relocating polluting production sites to developing countries; and (ii) developing country firms increasing their market shares in OECD countries at the expense of firms in these countries. See also B. Bergsto and S. B. Endresen, 'From North to South: A Locational Shift in Industrial Pollution?', *Norwegian Journal of Geography*, vol. 46, no. 4 (December 1992).
3. To identify those industries which would have to incur relatively high adjustment costs in order to reduce their polluting effects, the present study uses data reported in the US Bureau of Census publication, 'Pollution Abatement Costs and Expenditures', Department of Commerce, Bureau of the Census, 1991. Products produced by these industries are referred to as products/exports from polluting industries, and the industries as polluting industries. The classification of these industries is also discussed in the fol-

lowing section of this chapter. See also UNCTAD, 'Sustainable Development: Trade and Environment – The Impact of Environment-related Policies on Export Competitiveness and Market Access', TD/B/41(1)/4 and TD/B/41(1)/4/Add 1 (Geneva: UNCTAD, 1994).

4. US Department of Commerce, Bureau of the Census, op. cit. In this context it is worth noting that FUNCEX (Foundation of External Trade) in Brazil has undertaken a study on the sensitivity of Brazilian exports to environmental factors based on Brazilian data. The study was undertaken as part of the UNCTAD/UNDP project: 'Reconciliation of Environmental and Trade Policies'. US data has been used in a few studies examining developing countries. For example Gomez-Lobo used data on toxic emissions derived from EPA's Toxic Release Inventory used in the study by Lucas (referred to in C. Pearson, 'Trade and Environment: The United States Experience', paper prepared for UNCTAD, December 1993), to measure the impact of trade liberalisation in Chile in the 1970s on the pollution intensity of the manufacturing industry. See A. Gomez-Lobo, 'Las consecuencias ambientales de la apertura comercial en Chile', Colleccion Estudios CIEPLAN 35, (Chile, September 1992). Gomez-Lobo found that the share of polluting industries diminished during the 1970s. Ten Kate used US data to assess trends in the pollution intensity of Mexican manufactured industries in an attempt to analyse whether Mexico's industrialisation policies had contributed to environmental degradation between 1950 and 1989. See A. Ten Kate, 'Industrial Development and the Environment in Mexico', Policy Research Paper 1125, World Bank (USA, April 1993). Ten Kate found that the average pollution intensity of Mexican industry had increased over the period under analysis, but added that this might have been the result of composition effects and factors other than the industrialisation policies.

5. See also note 2. The types of cost included are, for example, expenditure on sewage and waste disposal services and direct operating costs to firms for air and water pollution abatement and waste disposal. On the basis of this information, 109 4-digit SITC Revision 2 industries were identified, all of which allocate 2 per cent or more of the value of their total 1991 sales to pollution abatement costs (operating costs). It should also be noted that this data does not reflect all environmental costs. For example the costs of compliance for environmental zoning, delays from permitting requirements, as well as workplace health and safety standards are not included. It should also be noted that the methodology is based on a static analysis of technologies used in 1991. It is possible that since then polluting technologies have been replaced by clean(er) technologies and that for some industries the operating costs may have decreased. Moreover pollution abatement costs are sometimes classified as such without being directly motivated by concern for the environment, but by profit-maximising business behaviour. See also OECD, 'Pollution Abatement and Control Expenditure in OECD Countries', OECD Environment Monographs no. 75, OCDE/GD (93)91 (Paris: OECD, 1993).

6. See US Congress, Office of Technology Assessment, 'Industry, Technology and the Environment – Competitive Challenges and Business Opportunities', January 1994, pp. 187, 190.

7. Derived from O. J. Mandeng, 'International Competitiveness and Specialization', CEPAL Review no. 45 (Chile, December 1991).

8. Because of its importance the OECD market is used as a proxy for total world trade.
9. Figure 5.1 illustrates the distribution of these combinations in a matrix with each of the quadrants representing one of the four possible combinations mentioned above.
10. F. Fajnzylber, 'International Insertion and Institutional Renewal', *CEPAL Review*, no. 44 (August 1991).
11. UNCTAD, 1994, op. cit.

Part II

Country Case-Studies

6 Environmental and Trade Considerations in the South Asian Region

Prodipto Ghosh, Preeti Soni, M.C. Verma and Rakesh Shahani*

The South Asian region comprises Bangladesh, Bhutan, India, the Maldives, Nepal, Pakistan and Sri Lanka. These countries all share historical, geographical, cultural and economic links. They are all developing countries, their per capita income on average being less than US$650. To develop economically they will need to cooperate with each other and with countries outside the region in a range of policy-making areas, but especially in trade. Trade has been recognised by all countries of the region as a pivotal means of expanding the resource base, furnishing economies of scale, reducing information costs and leading to more cost-effective production.

To date policy makers have largely been unmindful of the environmental degradation that can be caused by development activity. However, with the rapid increase in environmental problems such as deforestation, water and air pollution, there has been a realisation that development policy must go hand in hand with environmental policy. This realisation was reinforced to a significant extent by the findings and prescriptions that emerged from the Rio Conference on Environment and Development.

Countries within the region have gradually increased their cooperation in diverse fields, largely under the umbrella of the South Asian Agreement on Regional Cooperation (SAARC). A significant new development in intraregional cooperation is the recent agreement to reduce intraregional trade barriers: the South Asian Preferential Trading Arrangement or SAPTA. While the scope for mutual benefit through trade expansion has been largely appreciated, concerns have been expressed, as indeed they have in other regional trading blocs, that the objectives of SAPTA may not be fully realised if member countries use domestic environmental regulations as non-tariff barriers to imports from other countries. On the other hand, increased trade has the potential to further degrade the environment unless strong environmental regulations are put in place.

This chapter has two principal purposes. First it presents a summary of the major environmental concerns of the South Asian region, and second it discusses the possibilities for regional cooperation in expanding trade.

AN OVERVIEW OF ENVIRONMENTAL CONCERNS IN THE SOUTH ASIAN REGION

The South Asian region is endowed with abundant natural resources, but it is also prone to massive unforeseen dislocations, such as hurricanes and floods, that are both natural and man-made. These factors, plus a rapidly increasing population and recent industrialisation, have led to considerable environmental degradation in recent years. Of particular concern is degradation of arable land, pollution of inland waterways, alteration of coastal and marine environments and increased pressure on urban environments.

Land

The South Asian region constitutes only 3.5 per cent of the total land area of the world, but supports around 20 per cent of the world's population, leading to intense population pressure on the land. The region's land ecosystem has consequently deteriorated over the years, primarily due to changes in land-use patterns, deforestation and land degradation.

Long-term patterns of land use have changed dramatically in recent years. The expansion of cropland, shifting agriculture, encroachment into forests and other natural habitat for development and construction activities and an increased dependence on forest resources are major causes for concern. For example in Sri Lanka, with tea and coffee plantations replacing rainforest-covered hill slopes, there has been a significant loss of forests, with erosion being only one of the resulting problems. Certain mangrove swamps and wetlands have also been cleared for aquaculture farms, which has reduced the natural productivity and cleansing function of these highly sensitive ecosystems. Similarly, in Nepal increasing demand for agricultural land has resulted in widespread encroachment into forest lands.

Deforestation is a serious concern in the South Asian region. Although the region supports about 2 per cent of the total forest area of the world, forest clearing for settlement and development activities, overharvesting of trees and increased dependence on forest resources for domestic use, fuelwood and trade has resulted in the rapid disappearance of forests. In terms of the annual rates of deforestation, South Asian countries can be divided into three categories representing high deforestation, low deforestation and no significant deforestation. Nepal, with 4 per cent of forests cut per year, and Sri Lanka, with 3.5 per cent of forests cut per year, can be placed in the first category. In fact Nepal has one of the highest annual rates of deforestation in the world and certainly has the highest rates within the Asia/Pacific region as a whole. Only 15 per cent of forested land in Nepal carries dense forests and only 1 per cent predominantly carries young generation or pole-sized trees. Average growing stock is now less than 100 cubic meters per hectare. Although low deforestation rates exist in India, there are still significant losses in forest area, with an estimated annual loss of around 47 500 hectares per annum. While the lowest rates of deforestation occur in Bangladesh, Bhutan and Pakistan, even in these countries there is cause for concern. For example in Bangladesh the forested area has declined from 15 per cent to 5 per cent of the total land area in just 20 years. An important example of the linkage between trade and the environment was the discontinuation of the earlier voluminous exports of timber from Nepal to India a decade ago because of the unsustainable nature of the large-scale logging that was occurring in Nepal.

Land degradation such as erosion, salinisation, waterlogging and desertification may occur because of unsustainable land-use patterns, including deforestation, overgrazing of pastures, inappropriate irrigation practices, mining, improper use of nitrogenous fertilisers, intensive cropping, absence of proper organic management, poor drainage and inadequate soil conservation. India has the largest area of degraded land in South Asia, with about 50 per cent of its total land area degraded, and 27 per cent seriously affected by erosion. In Pakistan 25 per cent of unirrigated cultivated land has been degraded into desert due to soil erosion, while 40 per cent of the irrigated land area is affected by salinisation. In Sri Lanka 23 per cent of irrigated land is also affected by salinisation.

Although the region supports a high level of diversity amongst its flora and fauna, the loss of wildlife habitat is also of serious concern, as is legal and illegal trade in wildlife. The main causes of lost wildlife habitat are encroachment by rural communities into areas of natural habitat, and pollution. Only in Bhutan does the remaining habitat exceed 50 per cent of the original land area. In Sri Lanka and the Maldives there is also serious concern about the trade in marine species and the degradation of coastal areas, including the filling in of wetlands and mangrove swamps as well as the mining of corals, which all lead to further losses of marine wildlife habitat.

Inland waterways

The quality of the inland waterways of South Asian countries has deteriorated seriously in recent years. This has mainly been due to pollution and excessive water withdrawals.

There are several factors contributing to the pollution of inland waterways, including domestic sewage, industrial development and intensive agricultural production. Domestic sewage is often released into waterways in an untreated or only partially treated fashion. The organic wastes and suspended solids pollute water bodies and threaten both human health and aquatic life. Around 68 per cent of the population of South Asian countries do not have access to safe drinking water and consequently the prevalence of water-borne diseases is extraordinarily high. Industrial development, particularly chemical- and agriculture-based industries, also contribute to water pollution. Few of these industries are equipped with pollution control measures, leading to the release of quite toxic substances, such as cadmium and mercury, into the river waters of the region. The emphasis on intensive agricultural production using large quantities of chemical fertilisers and pesticides has also resulted in the pollution of inland waterways.

The exploitation of water resources has increased significantly in recent years for domestic, industrial and agricultural purposes. The increasing demand for water has necessitated construction of a number of dams in South Asian countries, although few have been constructed with environmental safeguards in mind. Consequently there have been many problems of inundation, sedimentation, habitat change and land degradation. Ground water is also being overexploited. India currently uses around 38 per cent of its total utilisable groundwater resources, with water tables falling dramatically in certain areas. Pakistan has utilised more than a third of its available ground water. In the Maldives, groundwater assessments for Male demonstrate that the supply of water for human use has reached a critical point. The depth of the water lens has shrunk from a level of 21 metres in the early 1970s to a mere three metres today.

Wetlands are increasingly subject to pollution and sedimentation, and many have already been reclaimed. For example Bangladesh has one of the most biologically diverse and productive wetland environments in the world. While these produce nearly half a million tonnes of fish annually and provide direct employment to 1.2 million people, they are rapidly being converted into agricultural land or shrimp farms, or are facing siltation and destruction through flood control, drainage and irrigation projects. Although Bangladesh is most severely affected by wetland loss, Sri Lanka and Pakistan are also rapidly losing their wetlands.

Coastal environment

The coastal environment of the region is important for marine-based food resources, minerals, trade and tourism. But again, it has been severely affected by alterations to existing habitat and overexploitation of resources.

Many of the region's mangrove forests have been destroyed because of land reclamation, the construction of ponds for fish and prawn aquaculture, and the felling of trees for fuel. Coral reefs have similarly been destroyed, mainly through the extraction of construction materials, such as coral lime for the production of white cement. This is particularly serious in the Maldives. Coral reefs have also been damaged through siltation, oil deposition, the release of toxic industrial effluents and fishing operations. The destruction of these coastal habitats combined with overfishing has also caused a serious depletion of the marine resources of the region.

Marine environment

The marine environment has been affected by land-based activities such as the pollution of inland waterways, and by ocean-based activities such as oil pollution. Land-based sources of marine pollution, including domestic waste, industrial effluents, sediment and agricultural run-off, all contribute significantly to the degradation of the region's marine environment. The rivers in South Asia carry 1.6 billion tonnes of sediments annually into South Asian seas. In India around 35 cubic kilometres of sewage is released into the marine environment each year. Vast quantities of industrial effluent and agricultural residues that are drained into inland waters also finally end up in the marine environment.

Ocean-based oil slicks that originate from oil shipping or off-shore oil installations also appreciably damage the marine environment. The sea route that runs from the Arabian Sea to the foot of India, adjacent to Sri Lanka, and across the southern part of the Bay of Bengal is prone to oil spills that can dramatically damage the coastal marine environments of India, Sri Lanka and the Maldives.

Human habitat

The human environment is also suffering from pollution and itself affects other environments. Major concerns persist in the region over the lack of adequate housing and infrastructural facilities, such as water supply, sanitation and transportation. The increasing industrialisation of cities and townships, and increasing population are also becoming very serious. Air pollution has reached alarming levels in several South Asian countries. The major contributors to air pollution are transport, power generation and industry. With a projection that the number of vehicles in the South Asian region will double by the year 2000, and industrialisation will continue to expand, air pollution has become a major environmental concern. In terms of suspended particulate matter, Delhi is one of the most polluted cities in Asia. But apart from cities and townships, the rural environment too suffers from a lack of proper sanitation, drainage, medical facilities, safe drinking water and other amenities. The rural environment also suffers from population pressure and agri-chemical applications. Although tourism is an important industry in the region, its environmental impacts are becoming significant. Tourism not only accentuates pressures on scarce natural resources, but also adds to the problems of waste disposal.

Global warming

Global warming will have a serious impact on the region. Several species may become extinct. Human welfare will suffer from increased heat stress, changes in water resources, shifts in agricultural productivity and declining forestry resources. Climate change and the corresponding rise in sea level are likely to cause increased flooding, particularly if storm activity also increases. Wetlands and lowlands will be inundated, mangrove swamps and coral ecosystems will be damaged, there will be increased intrusion of saline water into land areas, and there will be increasing difficulties for the storage and disposal of sewage. All of the coastal countries of the region are listed on UNEP's list of the 27 most vulnerable countries to sea level rise. Bangladesh, Pakistan and the Maldives are listed among the 10 most vulnerable in the world.[1]

MEASURES TAKEN TO PROTECT THE ENVIRONMENT

The environmental concerns in the region have been translated into action. All countries in the region have formulated environmental policies, developed institutions and spread awareness of environmental concerns throughout their populace.

The national level

Countries within the region share global concerns for environmental degradation. As a result, various governmental institutions and legal measures have been established to protect the environment, prevent pollution and lead to sustainable development. A summary of the major governmental organisations engaged in environmental management in these countries is given in Appendix 6.1, while a brief comparative table of significant environmental legislation and policy is included as Appendix 6.2. It should be noted that much of environmental law, policy and institutions are undergoing rapid change.

Environmental law in South Asian countries generally comprises natural resource and land-use management regulation, environmental administration laws and pollution control legislation.[2]

All the countries of South Asia have some form of natural resource and land-use management legislation. Unfortunately however, most of this legislation is directed to economic exploitation of these resources, rather than their sustainable development. While it is also common to have laws establishing some form of environmental administration and some type of environmental impact assessment procedure, the rigorous monitoring and enforcement of these laws differs widely between countries. For example, although the Maldives, Pakistan and Sri Lanka have legislation mandating environmental impact assessment, Bangladesh, India and Nepal as yet have no formal legislation but have simply produced administrative directives for this purpose. Pollution control legislation and standards for waste emissions have also been passed by a number of countries, but their enforcement faces difficulties in shortages of labour, technology and finance.

Besides the environmental legislation noted above, South Asian countries have also made commitments to many multilateral agreements concerning the environment, such as the Montreal Protocol on Substances that Deplete the Ozone Layer, the UN Framework Convention on Climate Change and the Convention on Biological Diversity.

A point worth noting is that hardly any major environmental laws in the region have trade provisions. One exception is the Bangladesh Pesticide Ordinance (1971), which provides for regulation of the import and use of pesticides in order to ensure the safety of public health.

The regional level

Owing to their geographical links, transnational environmental spillovers from one country of the region to another are inevitable. Nevertheless, to alleviate these spillovers and to manage the regional environment, a more cooperative approach should be taken to solve the environmental problems of the region. Although a South Asia Cooperative Environment Programme (SACEP) exists at the regional level it has not been very successful, although recently more significant moves towards regional cooperation occurred when the 'environment' was identified as a major area of regional concern during the Third SAARC Summit meeting. Many non-governmental organisations (NGOs) are also active, some of the more prominent NGOs being the Asian Wetland Bureau, the Asian Forum of Environment Journalists and the Asia Pacific People's Environment Network. Along with undertaking various environmentally orientated projects, general environmental awareness is being promoted by a number of United Nations agencies, including the United Nations Environment Programme, the United Nations Development Programme, the World Health Organisation and their regional agencies.

TRADE COOPERATION IN THE REGION

A major area of regional cooperation is trade. With a total population of over 1100 million, which is 6.2 times that of the European Union, significant fossil fuel deposits, abundant mineral ores, over 250 million hectares of quality farmlands, 86 million hectares of forest cover and 460 cubic kilometres of renewable water resources, the South Asian region has enough natural and human resources to sustain high growth rates provided they are managed appropriately.

In addition, the South Asian countries have undergone major structural change and have liberalised their economies in recent years. In India and Pakistan the industrial sector is relatively developed, although in other countries it is at a nascent stage of development. In all of these countries, except for the Maldives, agriculture still occupies an important place in the economy, even though its share in GDP has declined in recent years. The share of the service sector in GDP, on the other hand, has increased.

In spite of the existing potential, intraregional trade between South Asian countries at present accounts for just 3 per cent of total world trade. The share of South Asian countries in the total imports of the region was only US$2 billion out of a total 38 billion in 1990. The countries in the region rely heavily on OECD industrial market economies for both imports and exports. Accordingly, there appears to be significant scope for the promotion of intraregional trade and strengthening of economic cooperation.

Prospects for trade cooperation

A SAARC study on trade, manufactures and services has identified actual and potential products that could be traded between the countries in the South Asian region.[3] A further study, sponsored by the Committee on Studies for Cooperation in Development in South Asia (CSCD), identified about 110 products of export interest and about 113 products of import interest to the countries of this region.[4] A comparative perspective of these products indicates that along with competitiveness, there exists a high degree of complementarity in regional demand for many products. This implies immense potential for expansion of trade within the region.

Institutional changes required

The SAARC Secretariat's study on the potential of cooperation in trade, manufactures and services[5] has identified a number of policy initiatives and institutional changes that will be necessary for increased cooperation in trade, manufactures and services between the countries of the region. These include the following.

Trade policy measures: according most favoured nation (MFN) status to each other; introducing product-by-product approaches to extending tariff concessions on products of mutual interest, subject to appropriate rules of origin; freeing and rolling back non-tariff measures; applying the principle of regional preferences to joint ventures and technology transfer projects; appointing a group to negotiate tariff structures and preferential duties within a mutually agreed time frame; and setting up long-term project contracts, as an instrument of enhancing trade cooperation and stimulating efficient production among the member countries.

Measures for disseminating trade information: setting up a regional network of trade and technology information centres; stimulating exchange of trade related publications such as economic and business magazines and journals; arranging frequent exchanges of trade delegations and organising periodic SAARC trade fairs in individual member countries by rotation; promoting infrastructural development, improving transport infrastructure, and transit facilities for land-locked countries such as Bhutan and Nepal and between other member countries; introducing regular shipping services within the region with reasonable freight rates; upgrading and modernising postal services within the region; and opening branches and promoting joint ventures between commercial banks of member countries in each others' country.

Harmonisation of documentation: harmonising customs procedures, documentation such as letters of credit, national arbitration laws and procedures; and relaxing curbs on travel, particularly with respect to business people.

Clearing, payments and trade financing: providing suitable lines of credit to stimulate demand for expansion of intraregional trade; increasing cooperation among existing trade financing institutions such as import export banks and commercial banks, expanding membership of the ACU; exploring counter-trade and off-shore trading possibilities.

Joint ventures and investment: networking existing development finance institutions to facilitate identification and funding of regional projects and joint ventures.

Cooperation in the service sector: setting up joint promotions and technology transfers in tourism; arranging meetings between consultancy organisations of the SAARC region; developing small-scale and cottage industries and handicrafts; cooperating in the field of information, technology and related software; and setting up joint ventures in research and development in areas of common interest.

Institutional measures: setting up a SAARC Chamber of Commerce and Industry, networking experts and institutions in member states engaged in study and research on trade and investment matters; networking state trading organisations and other public sector organisations with regard to information on trade, technology and consultancy matters; and setting up a high-level committee on economic cooperation to follow up action on the recommendations agreed.

Harmonisation of standards. One of the constraints on mutual trade and other business relations within the region is the variety of standards adopted by the different countries. Despite being neighbours, information about the various regulations and standards in the region is difficult to obtain. This lack of information with regard to the various product and process standards could act as a non-tariff barrier. There is a need to share information and evolve a harmonised system of standardisation that is consistent with the WTO. Adoption, wherever possible, of near common standards within the SAARC region could pave the way for an increase in mutual trade.

Subsequent to these recommendations a Committee on Economic Cooperation (CEC) was established, which has more recently formed an

Intergovernmental Group on Trade Liberalisation (IGG), consisting of experts from the seven SAARC nations. The objective of these groups is to expand intra-SAARC trade.

THE SAARC PREFERENTIAL TRADING ARRANGEMENT (SAPTA)

A recent development in the area of regional cooperation between the countries of the South Asian region is the South Asian Preferential Trading Arrangement or 'SAPTA'. Guided by the declared commitment of the SAARC member countries to liberalise trade in the region in a manner that will enable all countries to share the benefits of trade equitably, and will take account of the special needs of the least developed countries (LDCs), the IGG drafted the SAPTA agreement, which will be completed by 1997.

SAPTA is designed to help promote and sustain mutual trade and economic cooperation among the countries of the region. Each member country will be able to request preferential treatment for a particular product from another member country on a reciprocal basis. This preferential treatment will apply to tariffs, para-tariffs, non-tariff barriers, commitment in respect of long- and medium-term contracts, preference in government and public procurement, state trading, and buy-back arrangements.

With the emergence of economic blocs throughout the world and a perceptible shift in world power politics from security to economic considerations, SAPTA appears to be a step in the right direction for the countries of this region.

CONCLUSIONS: TRADE AND THE ENVIRONMENT WITHIN THE SAPTA CONTEXT

While SAPTA may lead to mutually beneficial economic relations for member countries by providing open access to their markets, there is concern that an increase in trade may also encourage an increase in the exploitation of natural resources. However liberalised trade can also foster greater efficiency and higher productivity, and may actually reduce natural resource overexploitation and pollution by encouraging the adoption and diffusion of environmental management skills and cleaner production technologies. It has to be remembered that the primary cause of environmental problems is not liberalised trade, but the failure of markets and governments to price environmental resources appropriately.

Trade policies are at best uncertain tools for environmental management, as they usually only indirectly influence the use of environmental resources. Using more direct instruments to deal with environmental problems would better combat concerns such as deforestation, soil erosion or industrial pollution. Thus trade liberalisation measures should be accompanied by better and stricter use of environmental policies.

The two main concerns with respect to trade and environment in the South Asian context are (1) the potential use of domestic environmental legislation and policy as non-tariff barriers, and (2) the fears of environmentalists that free trade may involve greater exploitation of natural resources and environmental degradation.

The first set of concerns needs to be explicitly kept in mind as the SAPTA framework evolves. While the GATT/WTO system has already evolved norms, dispute panel rulings, and practices to limit the potential use of domestic environmental policies and practices to act as barriers to trade, it should be kept in mind that slight amendments of WTO norms may become necessary in the South Asian context as specific concerns emerge over time.

The second set of concerns should not be seen as an argument against regional trade liberalisation, but as strengthening the case for the enforcement of more effective domestic environmental legislation. Although all SAARC countries have made a start in setting up their regulatory environmental frameworks, with some having more elaborate machinery in place than others, clearly all are still in a learning phase. This presents considerable scope for mutual cooperation under the SAARC umbrella. In addition, experience gained in operating different regulatory regimes should be shared through regional workshops and conferences, although this should by no means require harmonisation of all environmental standards within the region. Joint environmental monitoring is a useful starting point for cooperation inside the context of the SAPTA. Regional training programmes and the requisite regional scientific infrastructure to undertake monitoring should be established quickly.

APPENDIX 6.1

Summary of the major governmental organisations engaged in environmental management and protection in the South Asian region

Country	*Major government bodies*	*Other government bodies*
Bangladesh	Ministry of the Environment and Forests	Space Research and Remote Sensing Organisation, Water Development Board, Soil Research Institute
Bhutan	National Environment Committee	National Environment Secretariat
India	Ministry of Environment and Forests	Central Pollution Control Board, National Wastelands Development Board, Indian Council of Forest Research and Education, Forest Research Institute, Council of Scientific and Industrial Research, National Environmental Engineering Research Institute, Natural Resources Management Systems
Maldives	Ministry of Planning and Environment	National Commission for the Protection of the Environment
Nepal	Ministry of Forests and Environment	Research Centre for Applied Science and Technology, Royal Nepal Academy for Science and Technology, Council for Conservation of Nature and Cultural Resources, King Mahendra Trust for Natural Conservation
Pakistan	Environmental and Urban Affairs Division, Ministry of Housing and Works	Pakistan Environment Protection Agency
Sri Lanka	Ministry for Environment and Parliamentary Affairs	Central Environment Authority, Environmental Council, National Aquatic Resources Agency, Urban Development Authority

Source: SAARC (1992).

APPENDIX 6.2

Evironmental law and policy in the South Asian region – a comparison

Country	*Major legislation*	*Policy and implementation*
Bangladesh	Bangladesh Pollution Control Ordinance (1977). Other laws include: Forests Act (1927), Bangladesh Municipal Act (1932), Factories Act (1965)	Environmental impacts assessments (EIAs) not mandatory under law. No approved environment quality standards at present, except tentative standards for water quality and some industrial effluents, but draft standards are being prepared
Bhutan	Most laws relate to forests. Main legislation: Forest Act (1969)(currently being updated)	National Forests Policy revised in 1988. National policy statements on air and water use also exist
India	Environment Protection Act (1986), Water Act (1974), Air Act (1980), Wildlife Protection Act (1972), Environmental Action Plan (1993)	EIAs required, National Forest Policy revised in 1988. National Policy Statements on Water Use also exist. Some national ambient standards exist and more are being established
Maldives	Environmental legislation in 20 vital areas. Housing Code (1987)	EIAs mandatory for any development project, National Commission recommends policies
Nepal	Soil and Water Conservation Act (1982)	National Conservation Strategy proposed
Pakistan	Pakistan Environmental Protection Ordinance (1983). National Conservation Strategy (1992). Several sectoral laws on a number of specific issues	National Conservation Strategy in place
Sri Lanka	National Environment Act (1980), several other laws deal with various aspects of the environment. Fundamental Right under the Constitution	EIAs required for any projects likely to have a significant impact on the environment

Source: SAARC, 1992.

Notes

* Prodipto Ghosh and Preeti Soni are from the Tata Energy Research Institute, New Delhi. M. C. Verma and Rakesh Shahani are from the Indian Council for Research on International Economic Relations, New Delhi.

1. SAARC (1992a).
2. See Rana (1991).
3. SAARC (1991).
4. Mukherji (1992).
5. SAARC (1991).

References

Asian Development Outlook (1991) (Manila: Asian Development Bank).

Mukherji, I. N. (1992) *Regional Trade, Investment, and Economic Cooperation among South Asian Countries* (New Delhi: Indian Council for Research on International Relations, International Centre for Economic Growth and East West Centre, Honolulu, San Francisco).

Ministry of Planning and Environment, Republic of Maldives (1992) *National Report – Maldives.*

Rana, K. N. (1991) *Environment, Energy and Infrastructure* (Kuala Lumpur: Indian Council for Research on International Economic Relations, New Delhi, Asian and Pacific Development Centre).

SAARC (1991) *SAARC Study on Trade, Manufactures and Services* (Kathmandu: SAARC Secretariat).

SAARC (1992a) *Regional Study on the Causes and Consequences of Natural Disasters and the Protection and Preservation of the Environment* (Kathmandu: SAARC Secretariat).

SAARC (1992b) *The Regional Study on Greenhouse Effect and its Impact on the Region* (Kathmandu: SAARC Secretariat).

World Bank (1992) *World Development Report* (Washington DC: World Bank).

7 The Sustainable Development of Leather Industries in Bangladesh

Fasih Uddin Mahtab

Recent trends in leather production worldwide point to the transfer of production from developed nations to developing nations, where low-cost labour is a comparative advantage. Although Bangladesh has made modest moves to establish itself as an important producer of leather and leather goods, several important technological and environmental concerns have arisen. These require urgent attention on the part of the government and industry if Bangladesh is to rely on leather production as a major support for its sustainable development. These concerns include the supply of quality raw hides and skins, the need to modernise existing plants, intensification of leather and leather goods research and development and control of pollution caused in the manufacture and tanning of leather. This chapter seeks briefly to examine these issues and provides some recommendations for how these technological and environmental concerns can be incorporated into an overall policy of sustainable leather production.

LEATHER PRODUCTION IN BANGLADESH

As in other parts of the Indian subcontinent, leather has been produced in Bangladesh for thousands of years. From time immemorial, leather manufacturers have produced vegetable-tanned leather – a method that is still in use today for the manufacture of shoe soles and industrial leather. Following the invention of chrome tanning in the West about one hundred years ago, however, the leather industry was revolutionised all over the world. In Bangladesh the manufacture of leather had traditionally been dominated by a vegetable-tanned product sold in local markets as well as being exported to Pakistan, Turkey and Iran, where it was in high demand for the manufacture of 'Kabuli Chappals'. But by the mid-1960s the tanning of leather using chrome in the wet–blue manufacturing process had firmly established itself. By the end of the 1960s there were over 30 medium-sized to large, partly mechanised leather plants in Bangladesh processing hides and skins using the wet–blue process for export. However environmental and health concerns about this began to emerge and in the late 1980s the Bangladesh government declared that wet–blue production for export would be banned after 30 June 1990. Consequently there was an earnest move on the part of the leather industry to build the required capacity for producing finished leather rather than wet–blue. As a result there has been no export of wet–blue leather since 1989, and now the export of crust-finished leather is dominant.

There are 225 registered leather tanning and finishing plants in Bangladesh, as well as more than 5000 small, cottage and family-type units.[1] The annual processing output of all of these industries combined is over seven million square metres of finished leather of exportable standard per annum. A wide range of leather types are manufactured, including crust, finished, suede, garment, chamois, plastic, upholstery, and fancy leathers such as tanned lizard skins.

Along with variations in the size of firms, there are also corresponding variations between firms with respect to the technology they use in leather tanning and finishing. This can range from quite primitive to highly sophisticated techniques. The non-mechanised, small and cottage-type firms mainly still produce wet–blue leathers using hand-operated techniques, such as sun-drying, hand coating and hand spraying. The medium to large industrial plants, in contrast, use sophisticated techniques, machinery and equipment. Italy, Germany, former Czechoslovakia and the United Kingdom are the leading suppliers of modern leather production machinery and equipment, although locally made machinery, such as paddles, wooden drums and toggle driers are still used. Manual methods, using hand tools of local origin and foot-operated sewing machines of European and Asian origin are commonly used in the cottage plants for manufacturing footwear. In the mechanised sector, footwear machinery and equipment comes from predominantly European sources, although equipment from developing countries such as Taiwan, India and South Korea is used as well. In several cases, joint venture projects have developed using technology from firms located overseas and Bangladeshi raw materials and labour.

Before 1947 all raw hides and skins collected in Bangladesh were sent to the tanneries of Calcutta and Kanpur in India for processing rather than being processed in Bangladesh. This situation has changed, however, and almost all leather is now processed in Bangladesh itself for export. The industry is scattered throughout the country, although there are concentrations in major cities such as Dhaka, Chittagong, Khulna and Rajshahi. Apart from the multinational subsidiary, Bata Shoe Co. (Bangladesh) Ltd, which is capable of producing around 50 000 pairs of shoes per day, the quantity of leather shoes produced overall per day is less than 500 shoes for the larger firms. Many of the small, cottage and family type units produce only 6–24 pairs of shoes per day in domestic finish leathers using low-quality raw materials rejected from the larger leather manufacturing plants. Consequently the leather products subsector of Bangladesh is still far behind international standards. However, since 1990 Bangladesh has made a modest effort to increase its production and export of leather goods, apart from footwear. Today Bangladesh also produces leather garments, briefcases, travel bags, wallets, jewellery boxes and golf bags, to name a few.

Although there have been marked ups and downs in total leather exports, this sector is still an important foreign exchange earner for Bangladesh, accounting for about 10.7 per cent of the total. Exports comprise top-grade leathers, while the local market is supplied mainly with lower-grade and rejected finished leathers. Italy tops the export destination list, followed by Brazil, France, China, Japan, Hong Kong, Russia, Germany, the Netherlands, Bulgaria, the United Kingdom, Hungary, Spain, the United States and Taiwan.

GLOBAL FOOTWEAR MARKETS

The making of footwear is a relatively straightforward operation and inevitably attracts considerable interest from low-labour-cost countries such as Bangladesh that are seeking to industrialise. From 1978 to 1990 there was a significant shift of shoemaking on a global scale from developed to developing countries. For example in 1978 developed country economies accounted for around 24 per cent of world shoe making and developing countries 53 per cent, but by 1990 that share had changed to 18 per cent and 69 per cent respectively. The main benefactor of this shift was Asia, which increased its share of world shoemaking from around 40 per cent to 60 per cent over that time, while Eastern Europe, the former Soviet Union, Western Europe and North America saw their combined share reduce from 47 per cent to 31 per cent.

The major shares over this period increased in two groups of countries. Brazil, China, the Republic of Korea and Taiwan registered increases of over 100 million pairs, while Italy, Portugal, Thailand and Yugoslavia experienced increases of over 50 million pairs each.[2] There were also several intraregional movements. For example production in Europe shifted from north to south and in East Asia, from the Republic of Korea and Taiwan to China, Indonesia, Thailand and Vietnam, with Malaysia and the Philippines also making their presence felt. At the other end of the scale, the United States saw its share of shoe making decrease by over 200 million pairs during that period and France, Germany and the United Kingdom also registered significant decreases. Footwear production in Belgium, Denmark, Ireland and Sweden has virtually ceased. China is now the leading producer, with 2700 million pairs, followed by the former Soviet Union, with 820 million pairs, although Asia as a region continues to dominate world shoe production.

The shift in shoe manufacturing within Asia has shown a clear trend. Production of simpler styles of shoes has moved away from countries such as Hong Kong, the Republic of Korea and Taiwan to countries such as China, Indonesia, Thailand and Vietnam, as manufacturing gravitates to lower-labour-cost producers.

On the consumption side of the global market, Europe, the former Soviet Union and North and Central America are the major consuming regions, while Asia and Western Asia are minor consuming regions. In terms of per capita consumption of shoes, the highest levels can be found in Western Europe and North America.

ISSUES OF CONCERN RELATING TO THE PRODUCTION OF LEATHER IN BANGLADESH

Several important concerns have arisen with regard to the Bangladesh leather industry that require urgent attention on the part of the government and industry. They include the supply of raw materials, modernisation of existing plants, research and development and pollution control.

Supply of raw materials

The availability of hides and skins for the Bangladesh leather industry is estimated at 15 million square metres of cattle hides, buffalo hides, goat skins and sheep skins. However there is low growth in the domestic ruminant population and the indigenous production of raw hides. To meet the requirements of the Bangladesh leather industry, live animals are imported from neighbouring countries, but these are of low quality due to skin diseases, physical abuse and poor nutrition in live animals. In addition the methods of recovery from carcasses and slaughtered animals are primitive, there are faults in storage and transportation, and outdated processing techniques are being used.

Although improvements could be made by enhancing the health and welfare of the animals, introducing new technology and new training and the use of polyurethane treatment to upgrade inferior hides, there seems to be no immediate prospect of a significant improvement in raw material supplies. The inferior quality of raw hides used in Bangladeshi leather production may significantly affect the future of the leather industry.

Modernisation

The existing tanneries in Bangladesh require urgent modernisation. A study by Huq and Islam[3] indicates that locally made machinery and equipment is invariably the least-cost solution for modernisation of small and medium-sized plants, but in the move towards larger-scale production a number of advanced European techniques such as hydraulic shaving, vibratory staking and rotopress production, will need to be implemented. However, apart from importing machinery and equipment from European sources, Indian-made machinery and equipment should also be considered for a number of sub-processes. Two examples include the hydraulic press and the pin-wheel measuring machine used in leather production, both of which are made in India. This mixed use of technology would tend to show a higher rate of return and also a higher labour component than if only European technologies were used. This is important in an economy such as Bangladesh, where low-cost labour is an important element of comparative advantage.

Research and development

In Bangladesh, research and development directed at leather and leather goods production is very limited compared with other countries. Only one section of the Bangladesh Council of Scientific and Industrial Research in Dhaka is responsible for leather-related research. This section, in fact, has only two rooms for a research laboratory, employing four scientists, and is poorly funded. Consequently government-funded research findings over the past thirty years have proved unsatisfactory. While some other institutes are undertaking leather and leather goods research, for sustained development of the leather industry in Bangladesh a serious research and development programme will have to be established.

Pollution control for Bangladesh leather tanneries

Most of the tanning industries in Bangladesh are located at Hazaribagh in Dhaka and in Chittagong. While there is some concern about the pollution caused by isolated tanneries scattered throughout the country, the main problem has arisen from the concentration in Hazaribagh. Here, large quantities of organic wastes and chromium have been discharged into the Buri Ganga river without treatment, except for makeshift impoundments located in the floodplains, causing widespread pollution.

The waste flows from the tanning industry can be considerable and particularly odorous. They arise throughout the leather manufacturing process from drips and spills, clean-up, initial washing, dehairing, liming, bating, pickling, tanning, retanning, bleaching and colouring. The pollutants vary widely and generally consist of discrete solids such a bits of flesh, hair and manure, dehairing chemicals and soluble proteins, oils, fat and grease, inorganic mineral salts and chromium. The primary polluting effects arise from floating solid wastes, chromium and the biochemical oxygen demand (BOD) caused by rotting organic wastes.

The effluent standards recommended by the government require some elementary treatment of wastes prior to discharge. In particular, sedimentation is required as a minimum, followed by some degree of biological or chemical treatment. Operators of tanneries are encouraged to screen their waste water and to separate the waste flows from different operations within a plant. However much of the more technical treatment schemes will only be feasible for the larger industries located in Hazaribagh because of the land constraints, especially for sludge disposal, that exist in rural areas. The government's overall intent at present is to provide some incentive for second-stage waste treatment, while avoiding overly restrictive measures.

It has been recommended that small rural tanneries treat their waste discharges by simple screening and batch sedimentation processes that do not require significant financial outlays. Although these approaches probably do not release effluents that meet the standards, the location of these industries in rural areas and their small discharges cause only minor adverse impacts. Indeed, except in situations where these industries have adequate land to install anaerobic or facultative ponds, the industry owners will not be able to provide effective biological treatment for the waste discharges, and in fact should hardly be expected to do so.

Reductions in waste water volumes might also be achieved relatively cheaply, by using water conservation practices. However, because the leather industries have ample groundwater supplies, there is at present little incentive to practice conservation. Likewise, although chemical recovery is feasible in the leather tanning industry, the technology requires capital investment and technical knowledge that is currently outside the reach of the leather industry in Bangladesh.

The major pollution concern lies in Hazaribagh, where 210 tanneries out of the total of 225 registered tanneries in Bangladesh are located. Because of the high levels of pollution produced, the government is considering relocation of these industries to a more isolated industrial zone. For this problem to be resolved quickly, pollution control facilities should be installed.

CONCLUSIONS

The global market trends clearly indicate that the prospect of expanding production in leather and leather goods industries is very bright. Both the tanning and shoemaking industries have tended to gravitate towards low-labour-cost producers, as confirmed by the enormous upsurge of shoemaking in China. While Bangladesh also has conditions suitable for this industry and the potential to join the list of producers, it is at present an unknown quantity. Nevertheless, since 1991 Bangladesh has made a modest beginning in its export of leather and leather goods, although this trend will require a quick response to a number of important questions if it is to continue. The industry will require a more comprehensive and aggressive government policy, the importation of more quality hides and skins, comprehensive measures to improve the quality of domestic hides and skins, modernisation of existing tanneries, a firm research and development base and efforts to promote pollution control.

The emerging links between trade and the environment warrants closer examination in Bangladesh concerning the status of various leather production technologies and their relative impact on the environment. Industrial growth and successful international trade in a country such as Bangladesh will play a crucial role in development efforts, but only if these opportunities are combined with a judicious mix of technology and environmental policy.

The environmental concerns arising from the production of leather are not too serious yet in Bangladesh, with the magnitude of industrial development still small enough to allow government and industry to focus on individual environmental problem areas. Simultaneously, environmental management systems appropriate to the country's overall conditions should be developed. But these environmental regulatory efforts must also take into account the question of the relative international competitiveness of the Bangladeshi leather industry so that any environmental measures do not undermine the potential of this industry.

Notes

1. Nooruddin, M. and A. S. Dey (1993) 'The Bangladesh Leather', on file with the Bangladesh Agriculture University.
2. United Nations Industrial Development Organisation, *Industry and Development: Annual Report 1992–93* (Vienna: UNIDO, 1993)
3. Huq, M. M. and K. N. M. Islam, *Choice of Technology: Leather Manufacturing in Bangladesh* (Dhaka: University Press, 1990).

8 Trade and the Environment: A Perspective from Bhutan

Achyut Bhandari*

The external trade sector is very important for the development of the Bhutanese economy. As the domestic manufacturing sector is small and limited, most goods for consumption and investment have to be imported while exports help to finance these imports. Bhutan's imports have always surpassed exports, even though the latter rose rapidly in the second half of the 1980s. Because of a steady inflow of foreign aid, the overall balance of payments remains in surplus. Apart from the export of cash crops, a few major industries also contribute to the growth of the export sector. Electricity is by far the most important export, followed by timber and wood products, calcium carbide and cement. The contribution of the export sector to GDP is increasing steadily.

India remains the single most important trading partner for Bhutan, accounting for about 90 per cent of Bhutan's trade. Bhutan's trade with Bangladesh has also picked up since the late 1980s with the export of apples, oranges and other agricultural, forest and mineral products. Bhutan's other trading partners at present are very few and are limited to Japan, Singapore, Thailand, Germany and the Netherlands. Bhutan does not really have any experience in exporting to developed countries because of its small export base, which concentrates on low-value high-volume commodities.

In its drive towards trade diversification, however, Bhutan cannot ignore opportunities outside the region for niche markets in products such as traditional handicrafts, mushrooms, essential oils, flavourings and herbs.[1] Trade contacts will have to be made through the help of projects financed with bilateral and multilateral assistance. Bhutan's unpolluted environment may also prove to be a positive factor in locating markets for some of these products.

What does all this mean for making trade and the environment mutually compatible for Bhutan? The benefits of trade are only too apparent, based on Bhutan's brief experience and the lessons to be learnt from those countries in East and South-East Asia who have followed a policy of export-led growth through an open economy. While we are unlikely to see Bhutan open its economy fully, the positive effects of liberalisation in our neighbourhood, particularly in India, may influence future policies. Furthermore the goal of sustainable development does not appear to be incompatible with the objective of attaining a higher standard of living through trade. The rationale of sustainability is that eradication of poverty and improvement in living standards will promote, not hinder, sustainable development.

Bhutan's major trade contacts in the near future are likely to be within the South Asian region, and in this respect, solutions to any problem associated with trade and the environment will have to be found at the bilateral and regional levels. Existing trade agreements with Bangladesh and India provide for the protection of human, animal and plant life as well as safeguarding other interests that may be mutually agreed. At the regional level, environmental issues in the context of trade may be raised in the SAARC forum. While SAPTA does not deal with the question of trade and the environment at present, cooperation in environmental matters is an important SAARC activity.

Because the volume of Bhutanese exports is small when compared with other countries in the region, and as Bhutan is not a member of the WTO, it does not have much experience in the workings of the multilateral trading system. Nor has it so far faced any problems regarding environmental quality or packaging standards for its exports, except in a few isolated instances that relate to quality control. In a recent trial involving the export of apples to Bangladesh, cardboard boxes were used for packaging instead of traditional wooden boxes. The purpose of promoting the use of cardboard for packaging instead of wooden boxes was to prolong the shelf life of apples and decrease the use of timber in Bhutan. In addition, as cardboard packaging is used predominantly outside the region, there was consideration of encouraging exporters to move over to using a more universally accepted form of packaging. However there was some resistance to the change from wooden boxes to cardboard packaging from the apple traders within Bhutan and from importers in Bangladesh. This was not because they did not appreciate the environmental considerations involved in a shift from wooden boxes to cardboard, but generally because of the greater durability of wooden boxes and their ability to protect apples during their rough handling and transportation from Bhutan to Bangladesh. While a consensus is yet to emerge, it is hoped that the traders will eventually see the advantages of packaging in cardboard boxes.

There are fears, particularly among developing countries, that the stronger developed countries will use trade policies to enforce environmental protection measures suitable to their needs. They could raise several questions relating to the production processes and environmental standards of imported products from developing countries. In these cases, developing countries may face an unfair and inequitable situation in which their exports become less competitive. While taking all possible measures to improve their environmental conditions, weaker developing countries may have to resist pressures from their stronger trade partners to impose environmental conditions unilaterally. In this context, a clear framework will be needed for the conduct of trade that leads to sustainable development. Trade should be regarded as a positive factor in promoting environmental protection and not as a tool for protectionism.

Meanwhile it is important for those Bhutanese agencies, firms and individuals responsible for trade matters to be aware of developments at the regional and international level. Unfortunately information on the subject is not widely available in Bhutan. Nor do the people concerned have the facilities or resources to acquire the necessary skills to deal with the issue. Resources for undertaking studies on particular aspects of trade and the environment and to conduct market surveys for environmentally friendly products remain scarce. It is in these areas that international organisations can assist Bhutan.

CONCLUSIONS

Environmental protection is high on the agenda of economic development in Bhutan and various initiatives are under way. While the potential for Bhutanese trade lies in the export of electricity, timber and wood products, minerals and some agricultural products, the government must also concentrate on improving the quality of these products or the processes by which they are produced so as to reinforce the nexus between trade and the environment. Hydroelectric and mineral development may have to be accompanied by a strong reforestation programme.

Bhutan's major trading partner is India, followed by Bangladesh. Trade in the near future is likely to concentrate in the South Asian region, although efforts will be made to expand to other regions. Any problems associated with trade and the environment from Bhutan's perspective will therefore have to be tackled at the bilateral or regional level, while keeping in mind developments at the global level. As the issue of trade and the environment is new to Bhutan, further international assistance would be beneficial.

Notes

* The views expressed in this chapter are the author's own and do not necessarily reflect the views of the Royal Government of Bhutan.

1. See Manisha Aryal, 'The Trade in Himalayan Herbs', *Himal* (Kathmandu, Jan/Feb. 1993).

9 Making Trade and Environmental Policies Compatible: Further Considerations From India

R. C. Jhamtani*

There are a number of factors, set at different political, economic and environmental levels, that will influence the potential for conflict between India's trade objectives and the environmental impact of those objectives. These include the political processes and institutions that will fashion the international and domestic debate on trade and the environment, and the Indian government's ability to influence these debates. Others include various economic factors, such as India's balance of payments position, foreign exchange situation and composition of trade and liberalisation of the economy. These will also determine India's role in the resolution of trade and environment issues. In addition, the impact of industry on India's own internal environment, particularly where goods are being produced for export, will also play a role. Finally, the constraints of revenue and the transfer and use of environmentally sound technologies will also be important. Throughout this chapter attempts are made to identify potential environment and trade links in the context of economic liberalisation.

SOME FACTORS INFLUENCING THE POTENTIAL FOR CONFLICT BETWEEN TRADE AND THE ENVIRONMENT

It is possible to consider 'quality of the environment' as an integral element of the natural resource endowment of a country, while historically international trade has been considered an efficient allocator of these resource endowments, based on the theory of comparative advantage. For example, Malaysian tropical forests being harvested for the needs of the Japanese industry, American car batteries being recycled in Sao Paulo and the translocation of polluting industries are all examples of an 'efficient' allocation of resource endowments. Within this perspective, due to the critical life-supporting nature of the environment, there are serious dangers in treating this resource just like any other.

It has been said that trade need not take precedence over environmental policy, and vice versa. Trade concerns arise from aspirations for development and economic growth while environmental concerns are rooted in conservation. Thus the trade and environment debate exposes issues that are rooted deeply in ethics and international equity. With the process of economic liberalisation, the regulatory role of governments in the context of environmental protection has assumed even greater significance.

There are a number of factors, set at different political, economic and environmental levels, that influence the potential for conflict between India's trade objectives and the environmental impact of those objectives. Broadly speaking there are political processes, economic forces and environmental impacts to consider.

The World Trade Organisation (WTO) is one of the principal political forces that will act as an interface between trade policy making and environmental policy making. In addition the 'green' movement both within India and internationally will play a significant political role as trade and environment policy evolves. The gap between the environmental standards of significant importing countries and exporting developing countries, including the ISO 9000 standards as well as standards related to products, production processes, packaging and the labelling of traded commodities, will also form another set of political pressures. Finally, compliance with international environmental agreements such as the Montreal Protocol, to which India is a party, will also provide a link between environmental and trade policy.

Several economic forces, both from within India and from developed countries, will influence India's policy making at home and abroad. These include India's balance of payments position, foreign exchange situation, composition of trade, and the overall relative strengths and weaknesses of the country's economy. In terms of India's trade with developed countries, characteristics such as their social and political structure, their higher public awareness of environmental concerns and higher income levels will influence India's environmental policy making.

Environmentally clean technologies, energy-efficient production processes and the equipment and software to monitor these industries will also need to be made available at affordable prices in order to maintain the relative competitiveness of India's export industries as the transition is made to more environmentally sound production processes. Apart from capital equipment, there is also a need for the upskilling of Indian technicians, access to designs and project consultants, and the financial resources to support the initial investment.

Institutional mechanisms will also be crucial in ensuring the transfer of these technologies and assisting with any heavy transitional costs. To date, some Indian industries and development finance institutions have already put into operation attractive schemes for these purposes, but some kind of 'clean technology Bank' may also be required. It is as yet unclear who will benefit most from the investment made in environmentally sound technologies: the producer of the technologies in developed countries; the India-based industries purchasing the technology and hoping to produce environmentally sound products; the ultimate consumer; or the environment itself.

Industries that might be most affected by the linking of trade to environmental policy would appear to include those dealing with heavy metals, toxic pesticides, dyes, chemicals, extraction of ores, energy efficiency and greenhouse-gas-related activities. In particular, leather, textiles, commercial plantation and agriculture industries will face constraints. Ore extraction and finished or semi-finished metal and engineering industries could also be affected.

Attractive tourist locations throughout India will have to reconsider the adequacy of their facilities, including the provision of clean water, garbage disposal and sewage treatment. Tourism is an important foreign exchange earner and attention must be focused on any environmental problems.

POTENTIAL EFFECTS OF TRADE LIBERALISATION ON THE INDIAN ENVIRONMENT

Attempts to integrate trade and environment policy making aim to reconcile environmental with developmental goals. The outcome of the reconciliation of trade and environmental issues will be unintended accidents unless a deliberate design is used in the process of making the two compatible.

The Indian government's trade liberalisation policy is primarily intended to make the Indian economy internationally competitive by giving Indian industries and consumers unhindered access to international goods and services. The protective measures that have hitherto been enjoyed by Indian industries will disappear gradually. While this should undoubtedly make Indian industries more efficient and raise consumer welfare, the loss of income from tariff-based revenue previously collected by the central government may also be significant. In the past, tariff-based revenue has been an important component of India's tax base, and has been spent on its development plans. Assuming there will be no further increases in other taxes, reduction in revenue from tariffs could significantly hurt the cause of the environment in India. This is because certain key socioeconomic sectors in the economy, such as industry, education, health and physical infrastructure, will be the last to have their budgets reduced, at the expense of the environment budget. Consequently the expenditure axe will in practice fall first on environmental and forestry expenditure.

The forestry sector is perhaps a useful one to examine in this regard. On the one hand India does not permit timber to be exported from India, but on the other, import duties on forestry products have been brought down considerably in recent years, from around 50 per cent to around 15 per cent on raw materials and 40 per cent on finished goods, such as veneered sheets. The combination of these two trade policies, a ban on exports and a reduced import tariff, work in the same direction, reinforcing each other by depressing domestic prices. Two possible effects appear to arise from this situation. First, the resulting demand for forestry products, raw materials and finished goods increases, if the price is elastic. Secondly, by making imported products available at lower prices, the domestic producer's profit margin is reduced, and over time this should also reduce domestic production, depending on the relative level of profitability of investment in alternative portfolios.

While the combined effect of these two impacts has yet to be studied in detail, it is important to take into consideration illegal tree felling and the price elasticity of demand for finished goods such as paper. A further reduction in prices for timber may encourage inefficient use of these products and also reduce the incentive for recycling. Perhaps a more judicious policy would be to raise the sales tax and excise duty on paper products. This might then yield a 'double dividend' by discouraging inefficient use of paper and other forestry products on the one hand, while on the other hand providing additional resources that can be earmarked for the forestry/environment sector. Low raw-material prices not only encourage waste but also reduce the incentive for private producers to plant industrial forests. The net effect may be degradation of the forests and the wider environment.

While some parts of the forestry sector in India are internationally competitive, such as the pulp and paper industry, there are other parts that have for many years enjoyed undue protection, such as the panel industry, leaving them inefficient in their operations. While liberalisation of trade will make many industries more competitive and successful internationally, policy makers will need to consider the employment and social issues that will result from the closure of inefficient plants.

MECHANISMS THAT WILL BE NEEDED FOR THE TRANSFER OF ENVIRONMENTALLY SOUND TECHNOLOGIES

Different titles are being used to describe environmentally sound technology, including 'low and non-waste technologies', 'waste recycling technologies', 'residue utilisation technologies' and 'resource recovery technologies'. While these titles are useful, there is a need for a more descriptive definition of what constitutes an environmentally sound technology. If an holistic and integrated approach to the polluting effects of an industrial product is taken, then the full life cycle of the product from 'cradle to grave' will need to be examined. Concepts such as reduction, reuse and recycling are also useful, as is the shift away from concentrating on 'end of the pipe' solutions to treatment *ab initio*. It also seems clear that ultimately the most appropriate environmentally sound technology will have to use only renewable resources for raw materials and energy sources. Transformation processes will have to be based on benign biological processes, rather than highly toxic, high-pressure and energy-intensive man-made chemical processes. However, as yet there seems to be no consensus about which approaches to apply and we are left with some confusion about what actually constitutes an environmentally sound technology.

There will also be important practical considerations to take into account when considering transfer or implementation of environmentally sound technologies. For instance commercially proven technology will need to be made available from domestic and foreign sources. Conditions for the adoption of environmentally sound technology by industries or project formulation and design agencies will also need to be created. This may require either the use of economic incentives or regulatory mechanisms. Close and continuous monitoring and enforcement of standards will be necessary, as will environmental impact assessment'. Both government and industry will need to upskill their staff. Finally, feedback concerning progress and realignment strategies will need-to be employed.

Financial institutions, project formulation and engineering design organisations, and environmental regulatory agencies will require education about the implementation and use of environmentally sound technologies by industry. To date these institutions have concentrated on 'end-of-pipe' solutions to pollution problems. Exposure of these agencies to other available options and sensitisation to these issues also appears critical. For example there may be more traditional or biotechnological alternatives, which may not necessarily come from internationally reputable laboratories or multinational corporations. 'Karnal Technology', which is now used in sewage treatment, is but one example of a locally based technology. Ironically, engineering solutions are often preferred by project designers and consultants because of the link between their consultancy fees and capital costs. If the case of environmentally sound technology is to be given serious impetus, then perhaps consultancy fees should be based on energy conversion efficiencies and inversely on the generation of waste. In any case, the right incentives need to be offered to project designers.

Most comparative analysis of technology choices continues to be based on financial prices rather than shadow prices. Often, quantitative valuation of the impacts of development on environmental amenities is not even attempted. When environmental values are considered, they are often assessed only in vague qualitative terms. Consequently environmental considerations are not seriously or rigorously built into decision-making processes. Most importantly, the agencies responsible for monitoring the impact of industry on the environment and enforcing regulations where there are breaches, have not generated up-to-date information on the most important environmental parameters.

CONCLUSIONS

The acute scarcity of financial resources and the refusal by developed countries to transfer commercially proven environmentally sound technologies at prices that can be afforded by firms in developing countries are important constraints on resolving trade and environment conflicts, as illustrated by the difficulty of finding substitutes for chlorofluorocarbons and toxic pesticides. It is simply unjust for developed countries, who have only recently been enlightened to the environmental crisis, to now require developing countries to 'leapfrog' dirty industrial technologies to new environmentally sound technologies without assistance.

But reconciliation of trade and environment issues will not just require the transfer of technology and assistance. It will require a new set of goals to be set at the global level by which all countries can equally and equitably abide. These goals should be rooted in global ethics and international equity, and until they are agreed upon and firmly set into place, all peoples will remain on a course that is unsustainable and ultimately life-threatening.

Note

* The author wishes to thank Dr Prodipto Ghosh (TERI) and Shri Arvinder Sachdev (Planning Commission) for their assistance in the writing of this chapter. However the views reflected are the author's own and are not be attributed to the Forest Planning Commission.

10 Protection of the Environment, Trade and India's Leather Exports

Ashok Jha*

The rationale for international trade stems from distinctive national endowments that lead to differences in comparative advantage. Theoretically, countries sell goods whose production costs are lower than others and purchase goods they cannot produce as efficiently as others. However the significance that different countries assign to environmental costs in the production process varies greatly and is dependent on a number of factors. One of the principal determinants is the level of income. Generally, the higher the income, the greater the concern for the environment.

Environmental concerns and trade are not *ipso facto* contradictory. Since trade is a recognised facilitator for economic development, growth in trade should have a favourable impact on the environment through an increase in income. But there is a danger that in the guise of environmental protection, non-tariff barriers to trade may be erected by countries that are losing their comparative advantage. However, to use trade as an instrument for achieving environmental goals is likely to be a poor strategy as the costs of lost trade will probably outweigh any environmental gains. If the malady is lack of environmental consciousness or environmental degradation, the remedy lies not in throwing up tariff barriers but in squarely addressing environmental considerations.

This chapter seeks to examine the interrelationship between trade and the environment, and in particular discusses environmental regulations that affect the exportation of India's leather and leather products.

LINKAGES BETWEEN TRADE AND THE ENVIRONMENT

International trade often externalises the internal environmental conditions of the producing country. Since the priorities of developing countries tend to focus on faster economic growth, environmental costs are often not fully reflected in the export prices of their products. This is not unusual given the pressure of poverty and other social factors that developing countries have to contend with. If the alternatives are cleaner air or less poverty, there is really no choice. Interestingly, however, through this process developing countries are partly paying for the environmental protection costs of developed countries. As the environmental costs of developing countries are not reflected in their export prices, they are subsidising consumers in the developed world by bearing the costs of damage to their own health and destruction of their own environment.

This raises a number of important issues. Are countries with lower environmental standards indulging in unfair trade? Can they be accused of subsidising environmentally damaging industries to the extent of their environmental costs? What about differences in variables other than the environment, such as wages, social benefits and health costs? In any case, trade restrictions are hardly the answer. For instance the banning of timber imports from a country in order to conserve the environment may only succeed in driving down domestic prices, thereby increasing local demand and leading to a more rapid destruction of the forests. Such an environmentally induced trade restriction therefore only introduces inefficiencies in both production and consumption.

While the benefits of more open trade and investment policies are widely known, what is perhaps not so widely recognised is the fact that these policies often lead to greater global environmental conservation. This can take place either through the transfer of more environmentally sound technology or by diverting production to those countries where the production process is less harmful to the environment. Agricultural production is a case in point. If the high subsidies that agriculture receives in some developed countries were to be withdrawn, more production would shift to developing countries where there is much less use of chemical fertilisers and pesticides. Overall environmental damage would be considerably reduced.

While international trade law allows each country the right to establish its own standards of environmental protection, provided certain basic principles such as non-discrimination, transparency, national treatment and avoidance of unnecessary obstacles to trade are respected, we need to be careful to ensure that environmental standards are not used for protectionist purposes.

In developed countries the creation of environmental standards are spilling over into North–South relations. These standards affect developing countries in different ways. The most important concern is that exports from developing countries will have to comply with the higher environmental standards in developed countries. Legally binding international environmental agreements on matters such as global warming and ozone layer depletion will also force industries in the developing world to adjust to new international eco-standards.

Often regulations in developed countries requiring minimum product standards adversely affect the trade of developing countries, especially if the standards have been established to protect domestic industry. The recent regulations promulgated in some countries, notably Germany and France, requiring manufactures and distributors of products either to take back packaging material or to recycle it are cases in point. On the face of it, these regulations are trade-neutral. German firms, however, have been developing alternative packaging material for some time while firms overseas have not. It seems, therefore, that the regulations afford a degree of protection to the German manufacturer and have given a comparative advantage to the German packaging industry. Packaging requirements can result in high adjustment costs for exporters to these countries, particularly for developing-country exporters, and this is likely to have serious repercussions on their competitiveness.

Other countries have introduced environmental labelling schemes. The multiplicity of these labelling schemes and their differing requirements is causing confusion and transparency problems. It is vital that efforts be made to find international consensus on labelling requirements to avoid any unintended barriers or restrictions on trade. In this context, the proposal to institutionalise an 'eco-label' for leather and leather products has to be considered very carefully. The idea apparently is for an eco-label to be issued to those producers who follow environmentally acceptable production processes and waste management practices. This would give them a trade advantage over products from countries not applying environmentally friendly production processes. But such a step could give rise to protectionism by creating a non-tariff barrier. What constitutes an environmentally friendly production process is a definitional issue that may differ from country to country. What is expected in an efficient effluent treatment and waste management system, for example? What is ecologically sound production? The answers to these questions will differ depending on the environmental and social conditions of the countries involved. The application of an eco-label is likely to restrict trade in leather and leather goods.

Of further concern is the expense associated with the proposals to change 'end of pipe' production processes to 'clean' production processes.

Environmental issues can become internationalised in various ways. Environmental policies may affect international competitiveness, or there may be transborder spillovers of pollution from one country into another country or globally. One way of dealing with this is for countries to resort to protectionism of one form or another. Trade could be used as an inducement or as a punishment to make countries comply with safer environmental standards. However the flaw in taking unilateral action is that one country's standards are imposed on other countries that may have different needs or environmental resource endowments. It is better to negotiate multilateral agreements based on international consensus. This assumes even greater significance if we take note of the special conditions and developmental requirements of developing countries.

ENVIRONMENTAL CONSIDERATIONS IN THE INDIAN LEATHER INDUSTRY

International trade has not been a major contributor to India's GDP in the past, mainly because of inward-looking import substitution policies. Consequently India's share of world trade has been declining and is currently a little less than 0.5 per cent.

More recently, however, India has embarked on a new strategy that involves liberalisation of trade in order to achieve greater economic growth. As this policy takes hold, environmental considerations in other countries will play a more important role in directing India's own domestic environmental policies. One export sector in which this has already begun to occur is the leather industry.

India's leather industry contributes significantly to her exports. Recently there has been a structural transformation in the leather industry, with the focus shifting from exports of raw hides and skins to exports of more value added products, such as shoes. The industry has certain inherent advantages, including its large raw-material base and competitively priced labour. It also tends to be composed of mainly small-scale tanneries located throughout the country.

In December 1989 the German government decided to ban the use of the toxic fungicide pentachlorophenol (PCP). The German ban was quickly followed by restrictions in Denmark, Sweden and the United States, although each country required different tolerance limits. Unfortunately this chemical was being used extensively for tanning by the Indian leather industry, which came under severe pressure to change to other fungicides. This exposed three main problems: a lack of information about the restrictions in other countries, a lack of chemical testing facilities in India and a lack of substitute chemicals.

Indian exporters generally become aware of relevant environmental policies in OECD countries through information in media reports, from travelling in those countries and from participating in international trade fairs. Information is also provided by a range of industry and government agencies, such as the Indian Central Leather Research Institute. Although initially there was inadequate information about the restrictions on the use of PCPs in other countries, particularly among the smaller tanneries, by the time the Indian government imposed its own ban, the information gap was well on the way to being bridged.

Since PCP can be used in different stages of production, for example in the preservation of raw hides and skins, in dyes or in fat liquor during the finishing process, the setting up of adequate testing facilities is important. Fortunately a number of testing facilities have been established quite quickly with assistance from the Indo–German Export Promotion Project as well as the Central Leather Research Institute.

Although chemicals such as TCMTB and PCMC have been identified as effective substitutes for PCP, most Indian tanneries now use Busan 30, which has to be imported from Germany or the United States. There are at present no local companies producing the TCMTB or PCMC substitutes for PCP, although one Indian company has just recently commenced production of Busan 30. The cost of using the PCP substitutes is roughly ten times greater than using PCP. It also appears that other chemicals, such as the Benzidine dyes used in the production of leather and leather products, are likely to come under closer environmental scrutiny in future.

While the restrictions on the use of PCP and some other chemicals can be justified on the basis of the health hazards they pose to consumers in foreign markets, there are some measures that clearly overstep legitimate concerns. For instance, process-related requirements such as efficient effluent treatment in tanneries would be tantamount to a non-tariff barrier. While the Indian leather industry is addressing this problem by setting up common effluent treatment plants as well as relocating tanneries so that pollution is minimised, this is an expensive process that could spell financial disaster for some of the smaller tanneries. It also raises fundamental issues about domestic sovereignty and transborder control of the production process.

Although the recent focus on environmental issues has put a strain on the leather industry, there have been a few positive spin offs. Natural products such as biocides, oils and fats, vegetable dyes and protein binders have become more popular. Vegetable-tanned leather, for which India is well known, is likely to be in greater demand by green consumers. In each of the major clusters of tanneries common effluent treatment plants have been established.

India has received assistance in meeting these new environmental standards. To mitigate the negative impact of environmental regulations in the leather sector, the Indo–German Export Promotion Project assisted in setting up testing facilities for PCP and also disseminated information on PCP regulations and standardisation requirements for testing procedures. The leather industry has received assistance from the Dutch Government in setting up pilot effluent treatment plants and chrome recovery plants. The UNDP-funded National Leather Development Programme has also sought to address environmental issues, especially in the eastern region of India by, for example, promoting an integrated leather complex so that smaller tanneries could be relocated to one location in Calcutta.

CONCLUSIONS

So far the impact of environmental regulations in other countries on India's trade has not been significant, although this is unlikely to remain the case as trade comes to occupy a more central position in the Indian economy. Much will also depend on the approach and attitude of developed countries in handling environmental issues. If these issues are used as a smokescreen for ushering in protectionism because their comparative advantage has been eroded, we could be in for what has been loosely described as 'environmental imperialism'. The question is not paramountcy between environmental conservation and open trade. The issue is one of the recognition of the substantive linkage between the two. An open, non-discriminatory trading regime backed by WTO rules and provisions is a better system for protecting the environment than policies that impinge on free trade. The sooner the North realises that comparative advantage is ephemeral and that attempts to try to cling to it through protectionist environmental policies are not optimal solutions, the better the chances that trade and environment policies will be complementary rather than contradictory.

Note

* This chapter represents the author's opinions and does not reflect the Government of India's views on the subject.

11 The Impact of Environmental Standards and Regulations Set in Foreign Markets on India's Exports

Vasantha Bharucha*

The compatibility between trade and environment policies will be an important international issue during the 1990s. For instance, while trade policy analysts fear that environmental standards may become obstacles to international trade, environmental groups are worried that trade liberalisation will lead to a downwards harmonisation of environmental standards. From another perspective, there are those who believe that products which conform to high environmental standards may accrue a competitive advantage.

Pressure on Indian manufacturers to improve their environmental performance will also arise out of governmental obligations under international environmental agreements, such as the Montreal Protocol on ozone-depleting substances. For instance, India has agreed, as have many other developing countries, to phase out the production and use of CFCs by the year 2006.

The market for environmentally friendly goods is located mainly in the member countries of the OECD, where during the last few years consumers have begun to articulate strong environmental concerns. These concerns have been translated both into individual purchasing decisions and governmental regulations. Of particular relevance to India has been German concern about the chemicals used in the Indian leather industry. OECD member countries account for 70 per cent of international trade and are the primary destination for Third World exports. Consequently, any appreciable changes in environmental standards in these markets will affect the rest of the world, developing countries in particular.

Developing countries such as India are also concerned that the commitments signed at Marrakesh will be undermined by the emergence of new non-tariff barriers in areas such as the environment and labour.

The ability of developing countries to implement environmental standards set by foreign markets will be limited by financial and technical constraints. Because domestic demand for 'environmentally friendly' products is almost non-existent in most developing countries, any products developed will have to be exported, reducing their economies of scale and competitiveness.

Also of concern for developing countries are the multitude of different eco-labelling programmes that have recently been established in OECD member countries. A balance needs to be struck between the differing schemes and international guidelines established to provide for harmonisation or equivalency. Developing countries must be consulted in these processes.

While Indian exports will no doubt confront new problems relating to the environment, it will make economic sense for India to incur these trade-related environmental costs in the short term to ensure continued access to rich overseas markets. India, along with many other developing countries, has liberalised its economy and there can be no turning back to the days of protectionism and export substitution. There will, however, be a need for international funding to support the transition to environmentally clean production processes.

This chapter examines the impact of these new environmental measures on a selected range of India's exported products.[1] It includes a survey of the common impacts, the refrigeration and air conditioning industry's use of ozone-depleting substances, tea, dyes and intermediates, agricultural products and processed foods, marine products, and the leather and textile sector.

ENVIRONMENTAL ACTION TAKEN BY INDIA AND INTERNATIONAL TRADE

Historically, living styles in India have largely been environmentally sustainable. Indeed, many Indian people still live in ways that conform closely to the sustainable development model. Organic farming, the application of natural and vegetable dyes, the use of biodegradable materials (such as jute and wood) in manufacturing and packaging are examples of India's tradition of sensible environmental practice.

Recent developments in agriculture and industry as well as concerns for accelerated development, competition and productivity have, however, shifted Indian agriculture and industry away from the use of natural materials and tradition. These changes have, in recent years, brought with them considerable degradation of the environment.

Action taken to address India's environmental issues has, to date, largely focused on domestic sustainable development, rather than environmental concerns in foreign countries. For instance, benzidine dyes have been banned in India because of local concerns about their effects on health, rather than through overseas pressures. In the food and beverage sectors, action has again been taken because of local concerns. Pesticide use has been reduced, with hazardous pesticides such as Malathion, DDT, and BHC being banned or restricted. The treatment of effluent is now being taken seriously in the leather and dyestuff sectors. With regard to packaging, India continues to use biodegradable substances such as jute, bamboo baskets, gunny bags and cotton, although it is now facing pressures to convert to materials that are recyclable rather than biodegradable.

India has also launched a series of domestic environmental programmes. In 1991 a project costing $263.6 million began with the objective of preventing and alleviating environmental degradation caused by industrial operations within India. Although scheduled for completion in 1997, the project was concluded early and has been followed by a second phase costing $330 million. A major component of these projects is the disbursement of loans to individual firms for setting up pollution control equipment.

Although India's major environmental concerns involve issues arising at the national level, in recent years there have been pressures for change from overseas export markets. In the early 1980s, for example, salmonella contamination in marine products resulted in the detention of shipments, but also quick changes in production methods. Germany's recent emphasis on removing PCP from leather products has resulted in a dramatic decline in India's leather exports to Germany, and has simultaneously compelled manufacturers to adjust to new processes that do not use PCPs. While major concern about the effect of new environmental standards on export consignments has in the past been relatively limited, it is expected to grow in the future.

ENVIRONMENTAL STANDARDS AND MARKET ACCESS: COMMON IMPACTS

The competitiveness of firms in developing countries is likely to be impaired by the introduction of stringent environmental standards in OECD member-country markets. High capital costs may have to be incurred in making any necessary changes, and unlike in developed countries there is often no domestic demand for environmentally friendly products that can be used to offset the costs of this capital expenditure. Consequently, the ability of firms in developing countries to comply with higher environmental standards will be limited by their access to overseas markets as well as financial and technical assistance. Moreover, governments in developing countries will find it difficult to subsidise the development of environmentally sound production because of competing claims to clean drinking water, education and health.

The competitiveness of products from developing countries may also be negatively affected by the way in which environmental standards are set in OECD member countries. In most eco-labelling schemes, for instance, only a narrow range of products are able to comply with the new criteria. The selection of criteria can also be so demanding as to specify only one particular kind of technology or production process, which may be out of the reach of firms in developing countries.

For those developing countries that import raw materials, the environmental quality of the final product will ultimately depend on the quality of the raw materials themselves. Even if those countries from which these materials are imported also adopt universal eco-criteria, other administrative problems such as plant inspection, insufficient information and differential treatment will frustrate compliance with the standards set in third markets. Information about which technologies are eco-friendly and which are not, or who the providers of this technology are, is often unavailable or questionable. Given the import requirements for some products in OECD member countries, many firms in developing countries now face serious problems concerning the degree of eco-compliance of any particular imported raw material.

The insistence by some OECD member countries for on-site plant inspection by different certifying agencies is also posing serious difficulties for developing countries. A number of products exported by developing countries, for instance, are produced in the informal sector where sponsorship of on-site plant inspection is impractical. These difficulties have been compounded by the failure to involve representatives of developing countries in the process of writing eco-criteria.

International cooperation is required to find answers to these PPM related issues. Technological transfer as well as financial and technical assistance seems appropriate. Eco-standards should also be set at the international rather than the national level to avoid the emergence of trade distortions. In some instances, harmonisation or equivalency between standards being set in different countries should also be more thoroughly explored.

Eco-labelling programmes

There are now a considerable number of eco-labelling programmes worldwide. The German 'Blue Angel' is the most famous, although there are also eco-labels in countries such as New Zealand and Canada. The European Union too has an eco-label. Among developing countries, the Republic of Korea, Singapore and now India have developed their own eco-labels.[2]

Packaging

Eco-packaging has emerged as an important area of environmental policy with significant trade effects. OECD statistics, for instance, indicate that packaging accounts for 20.8 per cent of all waste, 2 per cent of gaseous emissions, 1.5 per cent of water consumption and 3.7 per cent of energy consumption.

Measures being taken to reduce packaging or make existing packaging more environmentally sound include changes to the packaging materials being used, recycled content provisions, product charges, deposit-refund systems and take-back obligations. A range of initiatives have recently been taken by developed countries, and the European Union is now considering a harmonised directive to reduce packaging waste throughout the EU.

The costs to Indian exporters of these changes in packaging laws are many, but primarily include:

1. *Administrative costs*: the costs of administering take-back obligations under the German Packaging Ordinance, for instance, are estimated by Indian leather footwear exporters to be 1 per cent of sales.
2. *Compliance costs*: the costs of changing the design of packaging and/or the materials.
3. *Transaction costs*: the costs of obtaining information about the changes and complying with the differing packaging requirements among various countries.

REFRIGERATION AND AIR CONDITIONING

Chlorofluorocarbons (CFCs) have been produced on an industrial scale since the 1930s. While they were initially used as refrigerants, their favourable properties, and in particular their non-flammability and low toxicity, means that they have since been used in a range of other applications. Today CFCs are utilised as solvents and cleaners, refrigerants in air-conditioning, blowing agents for plastic production and propellants in spray cans.

In tropical and developing countries, such as India, the use of CFCs assumes an important role. Food preservation as well as drug and pharmaceutical manufacture are critical uses of CFCs in these countries. Although the refrigeration industry is relatively new to India, it is achieving high growth. The development of horticulture, aquaculture and floriculture, especially for export, for instance, has fostered above-average growth in the refrigeration industry. In most cases in India, however, CFCs are used more for life-supporting rather than life-comfort purposes.

Mounting scientific evidence of stratospheric ozone depletion in the 1980s led to concern about the contribution CFCs were making to the depletion of the ozone layer. In response, the international community concluded the 1985 Vienna Convention on the Protection of the Ozone Layer and the Montreal Protocol, which set timetables for the phasing out of CFCs and other ozone-depleting substances.

Under the Montreal Protocol, India has to phase out its production and consumption of CFCs by the year 2006. Since India is a very small per capita consumer of CFCs, it has been granted a grace period of ten years in which to arrange domestic compliance. Regarding exports to industrialised countries, however, CFCs will have to be phased out by 1996 or earlier. Indian exports to some countries in the Middle East, who have not signed the Protocol, will also have to stop in accordance with the provisions requiring a ban on exports to non-signatory nations. Fortunately India qualifies for financial aid and technical assistance, including the transfer of technologies, provided for by the Montreal Fund established under the Protocol.

Global production of CFCs was around 1 600 000 tonnes in 1990. Annual per capita consumption of CFCs was highest in the United States at 1.22 kg, with consumption in Europe and Japan reaching 0.93 kg each. India, on the other hand, consumed 0.005 kg of CFC per capita.

Five manufacturers produce around 20 000 tonnes of CFCs in India per annum. This may soon increase to 30 000 tonnes through bottlenecking. India is one of the very few developing countries to be self-sufficient in the production of CFCs and CFC based appliances. However, while less than 10 per cent of CFC production is used for refrigeration purposes in developed countries, India used 81 per cent of its CFCs for refrigeration and air conditioning. Unfortunately India has invested heavily in the CFC industry in recent years, particularly in the purchase of imported technologies, and many of the manufacturing plants are very new and have yet to complete their payback period.

CFCs are utilised in several ways in the Indian refrigeration and air-conditioning sector. They are used in domestic and industrial refrigerators and refrigerated cabinets, in water coolers, chillers, cold stores, transport and ice candy machines and also in the air-conditioning systems of homes and businesses, as well as automobiles, trains and shipping.

The phasing out of CFCs in India

A task force established by the government has assessed the incremental costs involved in phasing out controlled substances under the Montreal Protocol at $1400 million for an early phase-out and $2450 million for a late phase-out (that is, by 2010).[3] Clearly the costs of an early phase-out are substantially less than the those of a late phase-out. It seems crucial then for adherence to the Montreal Protocol to begin immediately. No new units for the manufacture of CFCs should be established and existing units should be phased out by 2006. New investment in non-CFC technology is urgently required as well as information about CFC-free technologies. In addition, the phase-out schedule for CFCs could be tightened, if additional funds supporting to meet the incremental costs are made available.

While the government may still be deliberating over what phase-out strategy to adopt, many Indian exporters have already shifted to substitutes in the knowledge that demand for CFC-free products in overseas markets is growing. Many Indian refrigeration and air conditioning manufacturers, for instance, have already begun the process of adjustment.

There is some confusion, however, about which substitutes should be adopted when replacing CFCs. While HCFCs and the new generation HFC134a are the preference of Northern industries, many environmentalists are concerned that these CFC replacements still have a negative impact on both the ozone layer and climate change. HFC134a, for example, is known to be 3 200 times more damaging than carbon dioxide when considering its contribution to climate change. According to environmentalists, a switch to hydrocarbons such as pentane and butane would be preferable to a wholesale switch to HFC134a. Indeed, the German company DKK Scharfenstein has already launched a refrigerator that operates on hydrocarbons. It appears to have been an overnight success with consumers. Many Indian manufacturers are very interested in this new technology, particularly as pentane and butane are freely available.[4] India's own research institutes are also examining possible substitutes for CFCs.[5]

It is clear that CFCs will need to be replaced. In some cases, however, the basic systems for switching over to substitutes are not presently available in India. It is also unfortunate that very few foreign companies have shown a desire to share their CFC-free technology. The adjustment costs to Indian industry, at the level of the individual firm, is estimated to be 30–35 per cent of current prices with respect to refrigerators and 10 per cent in terms of automobile air conditioners. These increased costs relate to technology, substitute raw materials and retooling.

TEA

Tea production is one of the oldest and most established industries in India and is of considerable importance to the national economy. India is responsible for nearly one third of global tea production, the remaining production taking place mainly in other developing countries in Asia. The markets of OECD member countries have become increasingly important for Indian tea exports, with about $1.10 billion worth of tea being exported to OECD countries each year.[6] OECD member countries, in particular the United States, Japan and members of the European Union account for 40 per cent of global imports. It is therefore vital that India not only retains these traditional markets for its teas, but also out-competes other nations in new markets.

It almost goes without saying, then, that any threat to India's overseas tea markets must be taken extremely seriously. In recent years concern has been growing about the level of pesticide residues in Indian tea. Germany, for instance, has made complaints about the high residue levels of ethion in Darjeeling teas. Complaints have also been received about Assam, Terai and Booras teas containing high levels of bicofol. In early 1994 there were fears that German tea importers would simply stop imports of Indian tea. There is no doubt that unless the Indian tea industry responds to these worries and ensures that the residue levels are reduced, exports of Indian tea will be seriously affected.

While the tea industry has taken some steps to deal with this problem, questions remain about whether these have been sufficient. The Indian Tea Research Association has issued a range of guidelines encouraging growers to take action to reduce the chemical residue content of their teas. These guidelines advocate spraying under proper supervision, spraying before plucking or spraying immediately after plucking, discarding the tea that is plucked immediately after spraying, the application of prophylactic treatments during the non-productive period ... rotation of chemicals and integrated pest management approaches. The guidelines and other information has been disseminated through various workshops and training programmes.

In addition, the government has banned the application of DDT, BHC, aldrin, aldrex, endrine, heptachlor, chlordane and tetradifon. Moreover, if chemicals such as thjiomton, dimethoate, malathion, moncrotopos, fenicypermethrin, fenvalerate, fluvalinate, phorat, phosphomidon, formothian, acephate, and carboxin are applied during the plucking season, the government's guidelines provide for discarding the plucking that immediately follows the spraying.

The Bureau of Indian Standards has selected tea as one of the food products eligible for the 'Eco-Mark' certificate, as long as certain conditions are complied with.[7] With an increasing domestic consumption of CTC tea, the export focus for tea should shift to Darjeeling tea and value added tea products, such as tea-bags and instant tea.

One difficulty facing Indian tea producers is that there is only one institute, the Pesticicles Residue Laboratory, that can test commercial samples of tea in India. It appears crucial for the Indian tea industry, along with important importing countries, such as Germany, to fund the establishment of further residue testing laboratories.

Another area of concern is the use of child labour in the tea plantations. Under the Indian Plantation Labour Act 1951, the employment of children aged 12–15 on tea estates is permissible. In tea factories, however, children must be aged 14 years or above. Nevertheless, the Child Labour Prohibition Act 1986 permits an exclusion for tea production because the limited use of child labour is not considered a social problem. The Indian tea industry is gradually discouraging employment of child labour in tea estates. Notwithstanding these legal provisions and their observance on the part of the tea industry, it is noted that children often accompany employed adults in the fields. Given the socioeconomic situation and the need for children to be with parents during the day, it appears difficult to find ways of avoiding this.

Indian teas have also been affected by the German Packaging Ordinance, which has required changes in the types of packaging materials used in the tea industry. Aluminum packs, for example, have been replaced by paper packs.

Recent increases in tea production arising from domestic and overseas demand has encouraged producers to clear more areas of forests for tea cultivation. The conversion of forest, however, has led to a range of environmental difficulties, including uncontrolled run-off and land slides. It seems appropriate for the government to legislate to prevent further expansion of tea cultivation and to require plantation owners to reforest existing estates. The expansion of acreage for tea envisaged under the Eighth Five Year Plan should be reconsidered.

DYES

The production of dyestuffs and pigments is a major function of the Indian chemical industry. Dyes are used as colorants in a variety of industries, including textiles, paper, leather, plastics and food. India has a well-established dyestuff industry. Moreover, in a transforming market where production is stagnant or shrinking in OECD member countries because of stringent pollution laws, India as well as other developing countries appear to have significant opportunities.

There are at present about 1000 units in India manufacturing various types of dye. These are located mainly in Bombay, Baroda, Surat, Vapi and Ankleshwar. This figure includes only about 50 units in the organised, large-scale sector. About one third of production takes place in the unorganised sector, of which about 60 per cent is exported.

The export of dyestuffs and their intermediates has risen rapidly, from Rs 400 million in 1980–81, to nearly Rs 3850 million in 1990–91. The United States and the former communist bloc countries form India's main markets for dyestuff exports. OECD countries account for nearly 60 per cent of India's exports of dyes and intermediates. China and Indonesia are the main competitors, with Japan, Korea, Taiwan, Singapore and Hong Kong also providing tough competition.

Eco-standards and India's response

There are no statutory eco-standards for dyestuffs. Nevertheless, European manufacturers of dyes and organic pigments have voluntarily formed the 'Ecological and Technological and Toxicological Association of Dyes and Organic Pigments Manufacturers' (ETAD) which recommends certain standards for handling, packaging and labelling dyestuffs.[8] Although ETAD is a voluntary organisation and does not have any authority to enforce compliance, it is compulsory for all ETAD members to adhere to ETAD's guidelines and standards.[9]

Voluntary standards are also being set by user industries. For example, eco-labelling has recently been introduced for textiles in Germany. The Eco-Tex Standard 100, or MST lists various criteria for evaluating textiles from an ecological perspective. Textile units in Europe are now seeking to ensure that their products fulfil the requirements of these eco-labels and are consequently exerting pressure on dyestuff manufacturers to ensure that their supplies meet the Eco-Tex specifications. Indian exporters of processed cloth and ready-made garments are also having to ensure that they comply with these eco-standards and are therefore insisting that their own domestic suppliers meet the same specifications.

India has banned benzidine-based dyes. Although the eco-standards for dyes and intermediates are very stringent in India (and in some cases more stringent than in developed countries) implementation and enforcement is inadequate. However, many organisations, such as the Bureau of Indian Standards (BIS), the Central Pollution Control Board and individual state boards, are specifying different standards and this is leading to non-uniform standards. Furthermore, in setting these standards no consideration has been made of the use of the best available technology at economical prices (BAT) or the best available process technology at economical prices (BPT). In some cases the eco-standards being implemented in India differ significantly from those being used abroad. Although industry generally seems aware of the need to comply with these eco-standards, there is a significant lack of awareness in the organised sector.

The dyestuff industry produces pollution in three forms: gaseous emissions, waste water discharges and solid wastes.

Operations in India's chemical plants differ dramatically between the organised and unorganised sectors. While many of the larger plants in the organised sector are now semi-automated and mechanised, those in the unorganised sector are usually operated manually. Nonetheless, even the most advanced units in India often still use antiquated machinery or processes such as tray dryers, which are now completely absent in the West. In contrast chemical plants in most developed countries are almost always fully mechanised and computerised, using the most advanced pollution control technology.

Many of India's outdated chemical plants are very polluting, hazardous to workers and inefficient. The picture in the unorganised sector is not very encouraging. Material handling systems are generally manual, and raw materials are often carried by workers on their backs. In the case of liquid chemicals, the transfer to reaction vessels is usually done by bucket, resulting in both spillage and harm to workers' skin. While safety norms are usually followed by the larger units, the levels of safety awareness are very low in the unorganised sector.[10] Again, while some primary and secondary effluent treatment is carried out by the larger units, in the unorganised sector the treatment of effluents is very basic.

A number of industries have requested government assistance in establishing common treatment plants in their chemical estates. While the smaller units are aware of their responsibilities with regard to pollution, the resources or technical skills for pollution abatement are often beyond their reach. In response, both state and central government have begun to fund common effluent treatment plants. It would also be useful if the government were to extend fiscal benefits for this kind of investment and expenditure.

Although pollution could be controlled using technology presently available in India, the costs are generally prohibitive, especially in the unorganised sector. There is also fear on the part of the organised sector that increases in their costs from pollution abatement expenditure will reduce their competitiveness compared with the unorganised sector. There is therefore a general reluctance to invest heavily in pollution control technology. In addition, while natural dyes may be preferred, their harvesting can be very labour-and energy-intensive as well as being logistically difficult.

AGRICULTURE AND MARINE PRODUCTS

Food safety is a major consumer concern in most developed countries, but particularly in the United States, Europe and Japan. Growing consumer consciousness about food and its qualities has encouraged organic farming and there is a rapidly developing market for organic produce in Europe and the United States. Raised consumer consciousness has also forced regulators to strengthen standards for pesticide residues, food additives and preservatives.

The food processing industry has attracted considerable attention in India because of its potential for earning foreign exchange and catalysing rural industrialisation. India is the second largest producer of vegetables in the world after China and the third largest producer of fruits. India has been promoting the establishment of modern food processing industries with the use of foreign equity and technology, encouraged by the availability of environmentally acceptable raw materials. It is also expected that conclusion of the Uruguay Round of GATT will improve market access for Indian agricultural products. Although the food processing sector has tremendous future potential, it is nevertheless at its nascent stage.

The impact of eco-standards on India's use of pesticides

Chemical pesticides have played an important role in increasing agricultural productivity in India, both in plant protection and post-harvesting processing and storage. But the use of pesticides has also had far-reaching side effects, particularly for the environment. Pesticides have affected the health of farmers and labourers and left behind residues in food, soils, rivers and ground water resources.

One feature of note in India is that while the use of pesticides in the production of food crops is prohibited, they are still permitted to be used as disinfectants in some health programmes. While DDT, for instance, cannot be used on crops, Hindustan Insecticides Ltd is permitted to produce 10 000 tonnes annually for the eradication of malaria. It is an unfortunate fact that the bulk of the DDT produced for these purposes is actually diverted for use on crops as it is one of the cheapest pesticides available, given the legal restrictions on its large-scale use.

With the conclusion of the Uruguay Round and the documents signed at Marrakesh, market access for Indian agro-products is projected to increase. It is therefore crucial that India's export enterprises meet the required consumer specifications in overseas markets before India's export advantage in this sector is eroded. Indian authorities are encouraging organic farming, which has significant appeal in Europe. The Agricultural and Processed Food Products and Export Development Authority (APEDA) has applied to the European Union for India to be registered as a source of organic produce. Information about organic farming is being disseminated and existing organic units are being registered by the APEDA. Apart from organic certification, the Indian Ministry of the Environment and Forests is developing food-product eco-mark criteria, particularly for edible oils, tea, coffee, beverages and baby food. However the concept of subscribing to ecolabelling schemes has not yet caught the fancy of Indian exporters, primarily because of a hesitancy to invest in the changes required to attain the ecolabels. Nonetheless this is expected to change in the near future as priorities shift towards more environmentally friendly products.

Seafood products

India is currently not meeting its potential as a manufacturer of quality seafood products. The Indian seafood industry by and large remains a supplier of raw materials to reprocessors in other countries, despite the significant value-added opportunities in this sector.

Unlike many other industries, environmental damage caused by the seafood industry tends to be localised in nature. Concerns have arisen about overexploitation, wasteful fishing methods, the production of effluent during fish processing and the overapplication of manure and other chemicals in aquaculture.

Marine products are considered an environmentally sensitive product group. Regulations imposed for sanitary and phyto-sanitary purposes concerning both products and production methods have in the past affected the market access of Indian products. For example salmonella contamination in the 1980s detained marine product shipments destined for Europe. Japan, too, has been particularly concerned with bacterial and toxin standards in seafood products. The United States presents special problems with consumer concerns related to the welfare of marine species such as turtles and marine mammals as well as sanitary issues. Although Indian seafood manufacturers have been careful to meet sanitary and phyto-sanitary requirements, they have not been so concerned with endangered-species issues. India is not, however, included in the list of 14 countries identified by the United States for the conservation of turtle resources.

Aquaculture has had mixed effects on the environment. Indiscriminate development of aquaculture ponds has adversely affected freshwater in many parts of the country, resulting in percolation of saline water into rice fields and agriculture lands. On the other hand aquaculture has helped to enrich the environment by creating large water bodies, such as Chilka Lake, that attract migratory birds. Ponds have also acted as buffer zones in coastal areas and stemmed the dependence of the poor on the forests.

THE IMPACT OF ECO-LABELLING SCHEMES ON INDIAN TEXTILES AND LEATHER, INCLUDING FOOTWEAR

Leather and footwear

Indian manufacturers have a range of fears about the eco-label criteria being established for leather products and footwear. These include a concern that increases in production costs will make those products manufactured in conformity with the criteria less competitive than non-labelled products. Another concern is that eco-labelling will provide an opportunity for animal right campaigners in the west to promote arguments against the use of leather. There are also the expenses incurred in the elimination of PCPs, laboratory testing,[11] physical and chemical analysis, adherence to German eco-packaging rules and the process of eco-label certification. Nevertheless products without an eco-label will no doubt face restriction in overseas markets.

Indian industry is very concerned about the implementation of eco-labelling in foreign markets, which has been taken without their participation and without technical or financial support being offered to assist developing countries to adapt. The failure of OECD member countries to adopt common standards in this area is frustrating for Indian manufacturers. The residue limits for PCP, for instance, differ markedly between Germany, Italy and France.

Indian Industry needs to work together to remove apprehensions about the environmental effects of manufacturing leather and footwear. Together with the scientific community, industry also needs to work on environmental criteria for eco-labelling and the adoption of eco-friendly technology in hide and skin processing and the manufacture of leather. Indeed, many eco-friendly techniques exist, but these need to be disseminated more widely.[12]

Textiles

For centuries India has been known for the quality of its textiles. The splendour of the clothes of the Maharajahs brought travellers to India from all over the world to trade in Indian fabrics. India has a massive textile base, with around 27 million yarn lengths, over 180 000 lakh looms, and about 1100 spinning and composite mills. Yarn production has reached 2000 million kilos per annum and cloth production has exceeded 200 000 million square metres.

The most serious environmental drawbacks in textile production occur during cotton growing and textile finishing. Cotton is one of the most chemical-intensive of all crops. The use of fertilisers and chemicals for growing cotton have eroded the soil and polluted water resources, as well as affecting the health of cotton pickers. While pollution caused by cotton growing is a major environmental concern, the processes used for textile finishing also cause serious pollution. Considerable quantities of water are used for bleaching, dyeing, mercerising, anti-felt finishing and water-proofing. For example around 100 litres of water are used in the production of just one kilogram of textile. The chemical contamination of textiles is also a health risk for workers and consumers.

Standards have been set for the pollution caused by the processing of cotton and these continue to be raised. The Indian Eco-Mark Scheme stipulates specific requirements for textiles in terms of their use of caustic soda and dermatologically safe materials in synthetics. Germany, for example, has now prohibited the importation of textiles or yarns that use dyes containing carcinogenic components or ingredients. The law, effective from 1 July 1995, prohibits the importation of any product printed or dyed with azo-dyes and dyes that contain or release trace quantities of nitrobenzene. The German authorities have begun to subject garments from India to extensive tests for traces of this chemical. It is expected that it will become mandatory for producers, suppliers and traders to provide a declaration that these chemicals are not present in their merchandise. The declaration will be binding and will allow German importers to decline to accept goods with traces of these chemicals without any legal recourse for the Indian exporter. Positive incentives are also required, however. Quota-free access for eco-friendly textiles should be considered. A scheme of joint-testing facilities to help support a number of small Indian companies should be considered.

The handloom textile industry in India is already using many methods that could be described as eco-friendly. Natural manure is used in the production of cotton. It is hand-picked and in the spinning and weaving process no chemicals are added. Although the production processes meet the required eco-standards set overseas, it will be necessary for an eco-mark to be applied before the product is acceptable in European Union countries. Rather than seeking eco-labels in an *ad hoc* way for each product, there have been suggestions that a national public relations campaign be run to promote Indian handloom textiles as naturally environmentally friendly or being based on biological inputs.

Cotton production and the processing of raw cotton is mainly undertaken by small farmers and cooperatives that are not yet aware of the eco-standards being applied to their product. Within the textile sector, exporters of readymade garments have the highest level of awareness because of their direct involvement with the rules in export markets and the expectations of foreign importers. Garment exporters, however, have been unable to implement the eco-standard requirements at the fabric manufacturer level because most of the organised mills concentrate on the domestic market, where environmental consciousness is limited, and because the decentralised sector is unaware of the need to adopt these new standards. Nevertheless, with the exception of the pesticide content of raw cotton, the composite mills could provide eco-friendly textiles if they were willing to do so.

The need to 'green' itself, literally from the ground up, is vital for the future of India's textile industry. While the industry has taken some small steps in this direction, support from both government and overseas will be necessary to make the large adjustments required.[13]

CONCLUSIONS

The issue of product- and process-related environmental standards is new to India. While tangible examples of the effects of these new standards are not always apparent, the potential for disruption to exports and the costs required for adjustment are of concern. At present the focus in India is on addressing national environmental concerns and attention has not generally been directed towards those issues specifically of a trade orientation. Nevertheless exporters are aware of eco-standards and the need for compliance. To facilitate the adjustment process, India's institutional framework for dealing with trade and environment issues will need to be strengthened. More information on compliance will be required regarding eco-standards and the eco-friendly technologies needed to comply with them.

Steps being taken by developing countries to accelerate their development seem to have focused on yield and productivity, with only an incidental concern for the environment. It seems that some polluting sectors have shifted from developed countries to the developing world, with exports in these sectors becoming an important developmental activity. Restrictions on imports for environmental purposes will retard this developmental process. Rather than penalising developing countries by imposing restrictions, developed nations should consider paying for the costs of adjustment, especially for those industries which have fled their borders. There is also some consensus that while environmental product standards might be prescribed for universal adoption by all countries, it is more desirable to leave the setting of process standards to national governments.

Developing countries face a dilemma. They have a choice between urgent budgetary claims relating to education, health and infrastructure or to the eco-concerns of foreigners that could, nonetheless, negatively affect their export and developmental activities. On the other hand, developing countries cannot ignore the environmental crisis because it has the potential to engulf them. The only sensible solution is to find the right balance between trade, development and the environment.

Notes and references

* This chapter presents the views of the author and does not necessarily reflect the views of the Government of India on the issues discussed.

1. This chapter is based on a study that examined all readily available reports/papers on the subject, with special reference to India, as well as a detailed questionnaire survey and discussions between the author and a number of industry representatives and the relevant institutions in every product sector.
2. The Bureau of Indian Standards, at the insistence of the Ministry of Environment and Forests, introduced a scheme for distinguishing and labelling environmentally friendly products with an 'Eco-Mark' label. The Central Pollution Control Board is the nodal agency for the Eco-Mark Scheme. The logo for the Eco-mark is an 'earthen pot', which represents the

earth and the fragility of its eco-systems and is a traditional Indian eco-friendly product widely known in the community.

The Steering Committee in the Ministry of Environment and Forests has identified 16 product categories for the development of Eco-Mark criteria: soaps and detergents, plastics paper, paints, electrical/electronic goods, packaging, food items, lubricating oils, textiles, cosmetics, aerosols, wood substitutes, food additives, batteries, pesticides, and drugs.

The general requirements for Eco-Mark certification include the listing of critical inputs, compliance with the existing provisions of various acts and regulations, reference to eco-friendliness in packaging and the use of recyclable or biodegradable materials in packaging.

3. It is worth noting that the task force estimates of these economic costs are significantly higher than estimates made by the World Bank (King and Munasinghe, study published by the World Bank and quoted in *Development Alternatives Newsletter*, 1991), whose estimates were only \$320 million for an early phase-out and \$482 million for a late phase-out, and the Ministry of Environment and Forest's consultants, Touche Ross, whose estimates were \$307 million for an early phase-out and \$703 million for a late phase-out. See K. Chatterjee 'After the Copenhagen Conference, a fresh look at the Montreal Protocol', in *Development Alternatives Newsletter*, vol. 3, no. 1 (1993).

There is, of course, a range of options for adjustment between the two extremes of early or late phase-out. An early phase-out would mean replacement of all CFCs as soon as possible, rather than using existing CFC capacity until the capital investment in the plants is defrayed. A late phase-out, on the other hand, would mean unconstrained growth of CFC production and consumption until the final phase-out date, which is at present 2010. The decision whether to follow an early or a late phase-out strategy affects the distribution of costs between producers and consumers of CFCs. With an early phase-out strategy the costs would be mainly carried by producers because their investment outlays in CFC technologies would not be fully recovered. With a late strategy, Indian producers would be better off, but the bulk of the adjustment costs would fall on consumers, who would face difficulties from for example, not being able to have their CFC refrigerators recharged after 2010. All studies reveal that when adding up the costs for all the three groups, a late phase-out strategy would be much more costly for India than an early phase-out strategy.

4. This refrigerator was the first to be awarded the environmental eco-label 'Blue Angel' by the Federal Office for the Environment (Germany). Seventy thousand of these refrigerators have been ordered by private households and retailers, despite production having not yet begun. The model works with a mixture of about 24 grams of propane and butane as a refrigerant. Propane and butane are natural gases available without licenses all over the world at prices of around 50 per cent of CFC prices and 10 per cent of HFC 134a prices. It is therefore particularly suitable for household refrigeration needs in the future growth markets of the developing world. The technology is easy to handle, requires no high-tech equipment or skill and creates no dependency on imports or expensive licenses. Although there have been some concerns over the flammable nature of these hydrocarbons,

in fact only 24 grames (the amount contained in two pocket lighters) is used. The United States Environmental Protection Agency has estimated that there would only be fifteen additional accidents each year if all American refrigerators ran on these hydrocarbons. The Indian refrigeration industry has expressed considerable interest in this new technology and is pursuing joint-venture proposals.

5. The institutes involved include the Indian Institute of Chemical Technology Hyderabad; the Indian Institute of Petroleum, Dehradun; the National Chemical Laboratory, Pune; the Indian Institute of Toxicological Research Centre, Lucknow; and the Centre for Mechanical Engineering Institute, Durgapur.
6. India exported about $100 million worth of tea to OECD member countries in 1993, which accounts for about 30 per cent of its total exports. Within the OECD, the United Kingdom accounted for 43 per cent of total imports, followed by Germany at 18 per cent, the United States at 16 per cent and Japan at 9 per cent.
7. Besides the criteria set as general requirements, tea must comply with certain product-specific requirements before it can attain an 'Eco-Mark' for example it must be free of adulterants (for example, spent tea leaves, grit, sand, and so on), have no odour or added colouring or flavouring, be free of mould and any pesticides residues must be within the limits specified. In addition, the iron content should not exceed 250 ppm, and the size of any iron particles should not exceed 2 ram. The lead content must not exceed 6.5 ppm.
8. The ETAD also conducts toxicological studies on various dyestuffs, maintains a data bank, prepares materials safety data sheets (MSDS) on hazardous chemicals, evaluates the limits of impurities tolerable in dyestuffs and issues guidelines thereon, and advises firms about the maximum allowable concentrations of various hazardous chemicals.
9. Nonetheless, ETAD has been instrumental in obtaining a ban in several countries on the manufacture and use of benzidine-based dyestuffs, which have proven to be carcinogenic. India has also banned the manufacture of benzidine based dyestuffs through notification 108(E), of 30 January 1993 (issued by the Ministry of Environment). Despite this notification there are still some manufacturers of these dyestuffs, including one firm in the organised sector in Gujarat as well as several cottage/small-scale units.
10. Most of these smaller units operate from a single shed, where restricted space does not allow the complete isolation of the hazardous processes involved. Even flammable liquids are often not separated or kept underground. Safety data sheets are not prepared by most small units and the concept of safety audits and risk analysis simply does not exist. Preventive maintenance, involving inspection, repairs and replacement of equipment on planned schedules, is not undertaken and in almost all cases replacement is made only after equipment actually breaks down.
11. Examples here are the charges for PCP tests levied by the Central Leather Research Institute (CLRD of approximately Rs 1500 per consignment of leather. This covers only the actual costs or special costs used in conducting PCP tests, such as ILR nitrogen, ILR oxygen, ILR helium of grade 1. If the additional costs of maintaining the equipment and the staff employed to

carry-out the tests were included, the costs would come to at least Rs 3000 per consignment tested for PCP in leather. These expenses need, of course, to be multiplied by the number of consignments undertaken for the PCP tests to arrive at an aggregate cost for just one test involved in the process required under the eco-criteria.

A rough estimate of the incremental costs that would be incurred in carrying out the basic tests set under the eco-criteria for leather appear to be 33 per cent of the present export price for leather footwear. Increases in costs of this magnitude will make Indian products uncompetitive, particularly when compared with shoes produced in China, Pakistan, Indonesia, Thailand and Bangladesh. Concern has also been raised about whether India's competitors will accept the same eco-standards.

12. The beam-house operation currently used, for example, is increasingly being replaced by eco-friendly methods. Enzymes are being used in other cases (these are non-toxic, odourless and biodegradable) instead of sodium sulphide in hair removing operations. In the processing of leather lining, ammonium chloride can be substituted by carbon dioxide. In the tanning of leather, technologies can be employed that recycle chromium sulphate, reducing the quantity of chromium used. The list of eco-friendlier processes is becoming more extensive.

13. Support was identified as important in the following areas:

    1. under the auspices of UNEP, technological upgrading and coordination with member organisations in other countries was necessary;
    2. replacing hazardous chemicals in the manufacture of dyes and auxiliary chemicals with suitable substitutes through the application of in-house research and/or modifying the technology;
    3. the use of safety alternatives while processing the textile items;
    4. the need for research organisations and testing agencies to establish reliable methods for trace analysis of chemicals;
    5. discouraging the manufacture of toxic chemicals;
    6. the creation of awareness about eco-standards through the various agencies.

12 Making Trade and Environmental Policy Making Mutually Compatible in Nepal

Khilendra N. Rana

The exports of Nepal mainly comprise items manufactured from natural fibres such as wool, cotton and jute as well as leather and a few kinds of agricultural produce. These exports may be affected by environmental legislation in OECD countries if it relates to the importation of these products, is directed at product and process standards for these items, or bans hazardous chemicals used in their manufacture.

The emerging consumer preference for 'green products' in OECD countries has important implications for Nepalese exports. In the West the wheel has turned full circle. As the harmful effects of chemical fertiliser and pesticide residues are revealed, there has been a shift from chemical to organic farming and 'environmentally friendly' or 'green' are the new catchwords. Products using conventional agricultural technologies or chemicals may be banned in the near future in OECD markets while green products are expected to fetch higher returns.

Although the need to feed an ever-increasing population in developing countries will require continued application of chemical fertilisers, pesticides and growth stimulators, the emerging consumer preferences in OECD countries clearly call for a switch to green products in the export sector. Such a shift will require a fundamental change in the agricultural practices of developing countries.[1]

The trade and environment interface has enabled OECD countries to impose many value-loaded conditions on the functioning of free trade, and these have influenced both production methods and consumption patterns. In other words, the environmental crusade has enabled the North to advocate non-tariff trade measures that have influenced environmental behaviour in the South. This chapter assesses Nepalese exporters' awareness of these emerging environmental regulations and their responses to them.

AWARENESS OF ENVIRONMENTAL REGULATIONS

It appears that most Nepalese exporters, while aware of developments in their own particular line of business, are not aware of the emerging overall trends in environmental policies and consumer preferences in OECD countries and generally have not taken any concrete action, individually or collectively, to safeguard their interests in this area.

While exporters have to comply with the environmental standards set by these countries, in the case of some export items the standards have not been clearly specified. This appears to be particularly the case for handmade products.

There is also no mechanism for a regular flow of information on environmental standards and regulations from OECD countries to Nepalese exporters. In addition Nepalese exporters do not have direct access to international organisations, nor is there any local channel for the flow of this information. So far no organisation in Nepal has been directed to monitor or analyse this information, and Nepalese exporters are guided solely by importing firms in OECD countries. This is probably because, as yet, Nepalese exporters have generally not experienced any serious difficulties in complying with the environmental standards of OECD countries, although some authorities have demanded certification from exporters. For example in 1993 Italian custom officials demanded certification that a leather consignment lying at an Italian port was the genuine hide of cattle and buffaloes and did not contain the hides of any wild animal. Similarly an importer of readymade garments in the United States recently demanded certification about the chemical composition of packaging materials.

Likewise there is no mechanism or authority in Nepal to coordinate trade and environment policies at the national level, nor any authority at the regional level. While environmental concerns have been raised in the SAARC forum, there is as yet no institution or officer to undertake coordination.[2] There is a vital need to educate Nepalese business people and government officials about the implications of trade and the environment. Hopefully assistance can be provided from regional and intergovernmental organisations such as UNCTAD, ITC and UNDP.

POSSIBLE IMPACTS

The emergence of environmental regulations in OECD countries may affect Nepalese export industries in a number of ways.

Although handmade export items such as woollen carpets and readymade garments are more sensitive to 'quality' and 'brandnames' rather than product standards, bans on the use of certain hazardous chemicals by OECD countries could profoundly affect these industries. The carpet industry uses large amounts of synthetic dyes as well as chemicals for carpet washing and moth-proofing. While the use of chemicals is not as extensive in the readymade garment industry, when Nepalese fabrics replace imported fabrics in the future the problem of chemicals may become more acute.

Perhaps the leather industry will be most affected by environmental regulations. Global attention has already focused on the use of pentachlorophenol (PCP) – a toxic fungicide used extensively in the leather industry. But leather chemicals such as ammonium and chromium salts, and benzidine-based dyes will almost certainly attract closer environmental attention in future. Eco-friendly saltless curing methods could be used in future to replace conventional processes that use non-biodegradable syntans and fat liquors, which presently cause significant pollution. Chromium tanning could be avoided altogether. It will become imperative for Nepalese industry to adopt clean technology in the leather sector so as to comply with the emerging eco-standards in OECD countries. This will require drastic changes to the use of existing technology and chemicals.

Packaging regulations may also affect the woollen carpet and readymade garment industries. If non-biodegradable plastics cannot be used for packaging, these industries will have to find some other biodegradable substitutes. While jute may be an alternative material, its properties must be improved for packaging purposes. Eco-labelling may also affect traditional Nepalese agricultural exports such as jute, pulses, vegetable oils and cakes, niger seeds and ginger. Luckily, sustainable agriculture based on permaculture methods has already gained a strong foothold in Nepal and could provide the basis for 'green' exports in the future.[3]

If the proposed carbon tax for the abatement of greenhouse gases in OECD countries is enforced globally, it will increase the price of fossil fuels and impose additional transportation costs on exports from and imports to landlocked countries such as Nepal. Preliminary analysis reveals that while a carbon tax may not have a significant impact on high-unit-value goods in Nepal, low-value bulky products may be seriously affected.[4]

However environmental concern and consumer preferences for green products in OECD countries could also impact favourably on exports from Nepal. Market opportunities may expand for conventional export products and new markets for environmentally friendly products may arise.

Although export items from Nepal are at present based primarily on organic or eco-friendly products, they can hardly be labelled 'green' on account of the use of synthetic chemicals in their processing. But this could be changed to suit consumer preferences in OECD markets. For example the handmade carpet industry could revert back to the use of vegetable dyes for wool dyeing, and with appropriate changes in washing and moth-proofing techniques, as well as packaging materials, the Nepalese carpet industry could truly turn out a green product. Similarly, the ready-made garment industry and leather industries could also adopt more environmentally friendly technologies.

Nepal might also look beyond traditional export items to expand its base of green products. Medicinal and aromatic plants, mushroom cultivation, vegetable seed production, flower seeds, vegetable production for export, fruit processing, tea development, sericulture and floriculture could be developed in an environmentally friendly way.[5] But the development of each type of new green product will need different lead times and involve different types of technical skill and know-how, including management and marketing.

Notes

1. A. T. Dudani, 'The Best Alternative, Naturally', *The Economic Times*, New Delhi, 25 July 1993.
2. SAARC, *Regional Study on the Causes and Consequences of Natural Disasters and the Protection and Preservation of the Environment* (Kathmandu: SAARC Secretariat, 1992).
3. Institute for Sustainable Agriculture, Nepal, *Newsletter*, vol. 7, no. 1 (Spring 1993).
4. The full analysis of the EU carbon tax proposal is not included in this chapter.
5. Ministry of Industry, HMG/N, 'Investment Opportunities: Feasible Project Areas', paper prepared for the Nepal Investment Forum, 30 Nov.–Dec. 1992.

References

Asian and Pacific Centre for Transfer of Technology (APCTT) (1992) *Hydrogen Study: Nepal* (Banglore: APCTT).

Environment Protection Council, HMG/N (1993) *Nepal Environmental Policy and Action Plan: Integrating Environment and Development* (Kathmandu: Environment Protection Council, August).

Fedesarrollo, O'Neil W. B. and D. Polania (1993) 'Trade and the Environment – Case Study Colombia: The Impact of a Carbon Tax on the Energy Sector', study prepared for the UNCTAD Secretariat, July.

Grubler A. (1993), 'Energy and Environment: Post-UNCED', in P. Ghosh and A. Jaitly (eds), *The Road from Rio* (New Delhi: Tata Energy Research Institute).

Sharma, C. K. and L. N. Bhattarai (1992) 'Sectoral Energy Demand of Nepal for the Year 1990/91', paper presented at ESCAP, Bangkok, 23–27 March.

UN (1993) *International Environmental Law: Emerging Trends and Implications for Transnational Corporations*, Environment Series, no. 3 (UN Department of Economic and Social Development, Transnational Corporation and Management Division).

13 Making Trade and Environmental Policy Making Mutually Compatible in Pakistan

Akhtar Hasan Khan

The environment has become a subject of great concern both at the national and the global level. It is multifaceted and has relationships with different spheres of economic activity such as agriculture, industry and ultimately trade. Having recognised this, Pakistan adopted a National Conservation Strategy in 1992 which aims to achieve conservation of national resources, sustainable development and improved efficiency in the use and management of resources. The operative part of this strategy includes programmes that will impinge upon Pakistan's exports. This chapter examines the impact of environmental concerns on the foreign trade of Pakistan. It also seeks to find ways of addressing these environmental concerns without distorting the free flow of foreign trade.

There are three central issues to be considered in making trade and environmental policy mutually compatible in Pakistan. The first is to identify any environmentally orientated import restrictions in countries that might hamper Pakistani exports. The second is to determine whether the aggressive pursuit of exports in Pakistan is leading to environmental degradation, and the third is to consider whether trade or tariff policy can be used to foster environmental ends.

ENVIRONMENTALLY ORIENTATED IMPORT RESTRICTIONS AND PAKISTAN'S EXPORTS

Pakistan, along with a number of other South Asian nations, may potentially suffer from the range of environmentally orientated policies being developed in OECD countries. These have been described generally in earlier chapters. However, of particular note are concerns about the inadequate flow of information to Pakistani exporters about environmental policies in other countries and the need to recognise opportunities in the 'green consumer' market.

It appears that Pakistani exporters are not fully aware of the relevant environmental policies being adopted in OECD countries, apart from general awareness of ISO 9000 and eco-labelling. Information about environmental standards and regulations is meant to flow to the export sector through multiple agencies, such as the Ministry of Environment and Urban Affairs, the Export Promotion Bureau, and Chambers of Commerce and Industry. However there does not appear to be an adequate supply of information and Pakistani exporters have not yet woken up to environmental concerns in importing countries. The multiple channels do not provide up-to-date information on this important subject. Nor is any systematic analysis of these concerns or mechanisms being sought, whereby Pakistani exporters could modify their products in order to obviate any difficulties or take advantage of new opportunities. Consequently there is an urgent need to set up a trade-related environmental information clearing house, either in the Export Promotion Bureau or in the Chambers of Commerce and Industry. Similarly, to date there has been little coordination of trade and environment policy between policy-making agencies in Pakistan. Trade policies are defined and implemented by the Federal Ministry of Commerce, while environmental policies are developed by the Ministry of Environment and Urban Affairs, but as yet there has been no compulsion on these two ministries to coordinate their activities in the area of trade and the environment.

One reason may be that few of Pakistan's exports have been affected by environmental standards, packaging regulations, eco-labelling, or the unilateral trade measures being applied by the United States. According to the Export Promotion Bureau, few major complaints have been registered regarding compliance with the environmental standards being enforced by major importing countries. The only major complaint received concerned the use of PCP, which was subsequently banned in Pakistan.

Many Pakistani exports consist of environmentally friendly products and exporters need to be made fully aware of the market opportunities provided by emerging environmental concerns. Exporters will need to be educated about the market niches for new environmentally friendly products, which they can exploit in order to cater for green consumers in OECD countries.

However Pakistan is starting to export 'eco-cotton' grown without pesticides. Eco-cotton enjoys a premium over ordinary cotton in the European market and enterprising Pakistani farmers are entering this field. Some

Pakistani exports are also being substantially increased by eco-labelling, such as the use of the Green Dot system in Germany. Where possible, Pakistani exports of leather and textiles are being certified as not using banned chemicals in their production. Exports would also be enhanced if exporters produced and packaged goods in a manner that complied with higher environmental standards.

Pakistan and other South Asian countries will need to accelerate their export performance if they are to compete with the countries of East Asia. To achieve this, production for export will need to become more diversified and move to higher value added export products. In particular, Pakistan needs to rationalise its cotton sector so that a higher proportion of the cotton exported is value added. If this can be achieved, Pakistan's agriculture base could be diversified away from cotton production, reducing pressure for land and water resources. Moreover, if long-term sustainable goals are to be realised, such as increasing Pakistan's forested areas, emphasis will have to be placed on using the land in more productive and value added ways. Production of high value added agricultural crops, such as fruit, vegetables, cut flowers, herbs and tropical plants will need to be encouraged, as will production of other eco-products. However it is important to note that promotion of these eco-products will require considerable attention to quality control and health standards, and that there will be a trade-off between the higher productivity of traditional goods and eco-products, which must be compensated for by improving quality and developing market niches.

PAKISTAN'S EXPORTS AND THE DOMESTIC ENVIRONMENT

Pakistan's export profile is dominated by textiles, cotton and rice. The cotton and textiles group, from raw cotton to readymade garments, accounts for 60 per cent of Pakistan's total exports. The leather group, which consists mainly of finished leather products, accounts for another 9 per cent, while rice (Basmati and Irri) accounts for about 5 per cent. The remaining 25 per cent of exports consist of miscellaneous items, in which carpets and fish are prominent.

Industrial environmental effects have generally not emerged as serious issues in national policy making, although some environmental concern is being expressed about the leather tanning industry, fisheries and the use of fertilisers in the cotton industry.

Of chief concern is the pollution that arises from slaughterhouses and tanneries in the leather industry. Land adjoining many tanneries has been contaminated by ground and surface water, and even where waste water is treated, the disposal of sludge remains a problem. The smell of rotting waste and flesh strapping can also be repulsive for people living in nearby areas. In addition there have been minor complaints over the use of PCP in the leather industry, but this has been resolved by turning to substitute chemicals.

In the case of fisheries, the major environmental concern is overexploitation and consequent stock depletion and loss of marine biodiversity. There has also been concern that aquaculture farms are damaging coastal ecosystems, especially mangrove swamps.

The use of synthetic fertilisers and chemical pesticides are also endemic problems in modern agriculture, but their use is essential for boosting agricultural output. Cotton farming has led to pesticide residues in the environment and pollution of lakes and rivers. The cotton and textile industries also use dyes that, when disposed of, can be toxic.

The use of pesticides in Pakistan has harmed predators as well as parasites, disrupting the natural cycle of pest management and control. While imported pesticides result in an immediate boost of output, they are harmful in the long run. The answer lies in integrated pest management, which is being trialed on a large scale. Similarly, some of the chemicals used in different industries cannot be easily drained after use and their residues pollute soils, the atmosphere and water supplies. The environmental hazards of different chemicals and the methods used for draining residues from waste water are not well understood in Pakistan.

THE USE OF TRADE OR TARIFF POLICY TO FOSTER ENVIRONMENTAL ENDS

Pakistan has begun to apply trade or tariff policies in order to foster better environmental outcomes. For example Pakistan has allowed timber to be imported without any duties or taxes. This has reduced the shortage of timber and the high prices that were leading to unsustainable cutting of local forests. Cheap imported wood appears to be reducing the inducement to cut trees for firewood. Differential tariffs have also been used to encourage the importation of environmentally friendly products. Imported goods required for environmental projects are being allowed into Pakistan free of customs duty in order to reduce the cost of such projects and improve their economic viability.

If there was greater knowledge about the environmental hazards of different chemicals and the options available for environmentally friendly substitutes, it might also be possible for Pakistan to determine more easily which products should be imported and which should not, or where the imposition of a differential tariff would be appropriate.

THE SCOPE FOR TECHNICAL ASSISTANCE

Pakistan requires technical assistance for the promotion of environmentally friendly production in the fields of agriculture, livestock management, fisheries, forestry and other industrial products. Technical assistance should, in the first instance, be directed at informing exporters about the environmental concerns held by OECD countries and the emerging relationship between trade and the environment at the international level. Second, exporters should be informed about the environmental standards being adopted in OECD countries regarding different products and how to go about conforming with them. In the third place, technical assistance should support exporters in producing environmentally friendly goods and services and help the government to develop and enhance certification institutions. This type of assistance could come from a range of institutions, including UNCTAD, UNDP, ITC, UNEP, ESCAP and so forth.

CONCLUSIONS

The most important consideration in the relationship between the environment and trade is that environmental concerns should not become disguised restrictions on international trade and give ammunition to the protectionist lobbies existing in all countries.

The lobby against the North American Free Trade Agreement in the United States has used the environment as one of its major opposing arguments. Because environmental standards in Mexico are much lower than in the United States, the lobbyists argued that industries in the United States can not compete on equal footing with their Mexican counterparts. They consider that some correction through trade is necessary to offset the higher costs in countries with higher environmental standards.

However these arguments are flawed. Environmental standards are a function of per capita income and become more stringent in countries where per capita income is greater than US$15 000 and there is an indifference in attitude between a higher income or a better environment. Moreover the developed countries have already passed through an earlier polluting stage of industrialisation and cannot now choke the development of latecomers by insisting on standards to which they did not conform themselves when they were at a similar stage. In the light of this, universal environmental standards should be opposed because environmental concerns vary between countries directly in relation to their various stages of economic development.

Finally, while there is an overlap between environment and trade policies, it occurs only in carefully selected cases. Trade policy can be used as a tool for promoting environmental ends where clear environmental targets are selected and trade measures are used in conjunction with other measures.

References

Anchanta, A. N., P. Dadhich *et al.* (1994) 'Requirements of Environmentally Sound Technologies (EST) to India for Compliance with Environmental Standards and Regulations in OECD Countries', paper prepared for the National Seminar on Trade and Environment, UNCTAD-TERI, 29 January 1994 (UNCTAD).

Government of Pakistan (1992) *National Conservation Strategy* (Government of Pakistan).

Government of Pakistan (1993) *Pakistan Foreign Trade Key Indications* (Government of Pakistan, November).

OECD (1992) 'Economic Instruments for Environmental Management in Developing Countries' (Paris: OECD, October).

Parikh *et al.* (1994) 'Tariff and Environment Linkages: A Case Study of India' (Indira Ghandi Institute of Development Research, January).

World Bank (1992) *International Trade and the Environment*, Patrick Low (ed.) (Washington DC: The World Bank).

14 Trade and the Environment: A Case Study from Sri Lanka

Lalith Heengama

With a population of 17 million people to support in a land area of only 16 million acres, Sri Lanka has one of the lowest per capita incomes in the world. In the past, industrial and agricultural practices such as timber felling, mining and fishing typically operated well within the absorptive capacities of Sri Lanka's different ecosystems. Trends over the past few decades, however, indicate that these traditional practices will no longer sustain Sri Lanka's burgeoning population. Although attempts have been made to enhance the natural resource base to serve this growing population, a variety of severe environmental problems have occurred.

From the Sri Lankan experience, it is clear that sustainable development will not be achieved if the very basis of this development is undermined. Protection of the environment and sustained use of natural resources are essential elements of development. Degrading soils, depleting resources and destroying ecosystems so as to raise incomes in the short term will inevitably undermine future productivity and the prospect of income for tomorrow.

THE CONCEPT OF SUSTAINABLE DEVELOPMENT

Development is commonly defined as the means by which the living standards of people are improved over time. The essential purpose of development is to improve education and health, equality of opportunity and generally to raise standards of living. Economic growth is a critical factor in achieving this sort of development.

Sustainable development adds a further dimension to the concept of development. The Brundtland Commission identified seven strategic factors required for sustainable development: (1) reviving growth, (2) changing the quality of growth, (3) meeting essential needs for employment, food, energy, water and sanitation, (4) ensuring a sustainable level of population, (5) conserving and enhancing the resource base, (6) reorientating technology and managing risk, and (7) merging environment and economics in decision-making.[1]

Therefore if development is to be sustainable it must be lasting and sustain those resources needed for future generations. Today's developers must take care not to cause excessive degradation of natural resources or pollute the environment. In this way, future generations will benefit from an inheritance of useful skills, education and knowledge as well as physical capital. Consequently a significant investment must be made today to preserve these resources for tomorrow.

TRADE LIBERALISATION AND ENVIRONMENTAL PROBLEMS

It is generally accepted that trade liberalisation does not have to conflict with sustainable development, provided certain measures are taken at both the national and the international level. However the experience of Sri Lanka and many other developing or newly industrialised countries has been one of unsustainable development, exacerbated by the liberalisation of their economies and international trade.

Sri Lanka's environmental problems, such as unsafe water supplies, inadequate sanitation, soil depletion and land degradation have been very similar to those encountered in other developing countries. These problems are now being added to, however, by new environmental problems associated with industrialisation. They include carbon dioxide emissions, photochemical smogs, acid rain, use and disposal of hazardous wastes, and emissions of ozone depleting substances.

Many of the industrial activities presently undertaken in Sri Lanka either commenced with or were expanded upon as a result of the adoption of an open economic policy in 1977.

One activity that developed very quickly was the tourist industry. In order to meet the growing tourist demand for accommodation, a large number of hotels and guest houses were constructed very quickly by foreign and local hoteliers in coastal locations without adequate impact assessment. Subsequently there have been severe erosion problems in many coastal areas, pollution from untreated waste disposal and complaints from local residents about restrictions on their access to the sea. Coastal areas have also suffered considerable degradation from mining, removal of sand and destruction of coral reefs. Despite efforts by the Sri Lankan authorities, it has been difficult to curb mining and sand removal because of the pressures of commercial and industrial development.

Another consequence of Sri Lanka's open economic policy has been a doubling of vehicle imports over the past five years. Often these vehicles are either used or reconditioned. Using imported Japanese engines, petrol driven vehicles are often converted to diesel to avoid the high cost of petrol and gain additional mileage. Many are in very poor condition. Vehicle inspectors are lax and the additional taxes that have recently been imposed seem to have had little effect in deterring this practice. Both the increased imports of vehicles and the use of diesel have contributed to large increases in lead and other atmospheric pollutants, particularly in towns and cities.

While chemicals have been used for agriculture and industry in Sri Lanka for a long time, their use has increased over recent years and with this a concern about their effect on the environment. With liberalisation of trade, a large variety of chemicals, including pesticides, herbicides, pharmaceuticals and industrial chemicals were imported into the country. The growth in the use of these chemicals has been such that now hardly any agricultural production is undertaken without their use. Unfortunately these chemicals have been used indiscriminantly and without education. Although attempts have been made to educate users, curtail the excessive use of chemicals and encourage users to revert to traditional methods of production, the sales promotion campaigns of the chemicals industry have been more effective and levels of chemical use have been maintained.

The excessive use of chemicals and the expansion of Sri Lanka's population has also resulted in a loss of the country's biological diversity. Urbanisation, tourism, mining, pollution and other industrial activities also threaten Sri Lanka's unique flora and fauna.

Sri Lankan forests too have suffered from increased economic activity. For instance a large proportion of Sri Lanka's forest area has been cleared to provide land for new village settlements, the expansion of existing villages, development schemes and agricultural production. Since independence in 1948, natural forests have been reduced by 50 per cent, with 3 per cent of the remaining forests being cut each year. Shifting cultivation, while acceptable in supporting a small population, has in recent years caused significant deforestation. Commercial forestry and firewood collection add to the list of problems. Although the government has made some efforts to reduce the impact on Sri Lanka's forests, for example by removing all customs duties on imported timber, forest exploitation continues to take place in an uncontrolled and unplanned manner.

Disposal of hazardous waste is becoming a significant problem in Sri Lanka. Many industries have no proper waste disposal facilities and their waste is often discharged directly into rivers and other waterways. There has been at least one attempt by a bogus exporter illegally to dump hazardous waste. Fortunately, in this case the port authorities discovered the waste before it was unloaded, and after much persuasion the waste was removed from Sri Lanka.

Although Sri Lanka has instituted environmental laws and, in particular, environmental impact assessment procedures, on some occasions the government has deliberately overridden these laws. For example, recently the government launched a programme to establish 200 garment factories in rural areas that were to be constructed in just one year. The purpose of these factories was to provide employment for rural youth and to earn much needed foreign exchange. Private developers were encouraged to build the factories on land that was either leased or acquired by the government. No environmental impact assessment was carried out. Unfortunately it seems that much of the land had been reserved for other purposes. In another example, the government decided to launch a project to build one million houses. Again the government chose to ignore its own environmental impact assessment procedures because of the drastic need to house many of the country's homeless. However there are also examples where projects have been subject to an assessment of their environmental impact. In one example, an application to mine gemstones in the Kaluganga River bed using mechanised equipment was refused on the ground that it would disturb the environment.

While the experience of most industrialised countries has been first to provide for growth in the economy and then to take action to protect the environment, this does not need to be the case. The difficulties experienced by other countries should act as an example for countries such as Sri Lanka that have only recently begun to industrialise. Environmental protection can and should be undertaken in concert with an open economic policy and trade liberalisation. Sri Lanka needs to learn these lessons and integrate the concepts of sustainable development more fully into economic planning.

TRADE LIBERALISATION AND EXPORTS

The report of the Bruntland Commission mentions that if developing countries are to reconcile their need for rapid export growth with their desire to conserve their natural resource base, it is imperative that they enjoy access to markets in industrialised countries and diversify their production base. Sri Lanka is one developing country that has recently tried to diversify its exports beyond the traditional commodities of tea, rubber and coconut in order to benefit from international trade. Indeed it is a noteworthy feature of the Sri Lankan economy that since the open economic policy was adopted, industrial exports have surpassed traditional agricultural exports as a proportion of total export earnings.

For example exported minerals have become a new and significant foreign exchange earner for the Sri Lankan economy. However, despite Sri Lanka being reasonably well endowed with minerals, there are concerns that these non-renewable exports will be quickly depleted.

Exporters need to be alert to new environmental policies in other countries and how these may affect their exports. It seems that most are unaware of those environmental standard specifications and regulations promulgated in OECD countries that will affect their industry, although certain recent examples have highlighted the problem. For instance the use of traditional packaging materials, such as tea chests, has caused difficulties for some exporters as they cannot be readily recycled. In another example, importers of desiccated coconut in OECD countries insisted that Sri Lankan suppliers use high-barrier coextruded film instead of polythene to package the coconut because some of the batches had been tainted by polythene odour. The garment industry – another significant export earner for Sri Lanka – was also affected by changes in packaging that meant local suppliers could not use wire to seal their packaging. As some of the packaging material that now needs to be used will have to be imported, some export orders will be delayed and extra costs will be incurred for the new packaging materials.

Eco-labelling schemes may also have adverse effects on Sri Lankan exports. It is feared that these types of environmental regulation will act as hidden tariff barriers because of the difficulties and expense that will be incurred by Sri Lankan manufacturers in obtaining eco-labels from foreign authorities. In future it will be necessary for exporters to establish before concluding any contracts that the packaging for a particular product is compatible with the waste management policies of the export market. Sri Lankan exporters will need to educate themselves about what these types of environmental policy will mean for them. Concepts such as 'reduce' 're-use' and 'recycle' will have to be applied in practice to their products, and especially to their product's packaging.

CONCLUSIONS

There is no question that trade and environment policies must be made mutually compatible, not least because both aim to increase welfare by efficiently allocating resources. Trading activities will not be sustainable unless they are compatible with the environment.

To bring about mutually compatible trade and environment policies, there must be a complete understanding between all countries engaged in international trade, whether exporting or importing, about what a mutually compatible policy would be. For example, while the WIDER Report (Study Group Series 9) on Indo-Sri Lanka Economic Cooperation highlighted the need for a bilateral agreement to facilitate trade expansion between India and Sri Lanka, there is also a need to incorporate environmental factors into this relationship. Since Sri Lanka and India are both members of GATT, any agreements concerning trade or trade and the environment should be consistent with their GATT rights and obligations.

There are also many lessons that developing countries could learn from the experiences of OECD countries. It has been proven that it is less expensive to prevent environmental degradation than to attempt to cure it after the fact. The costly clean-up of hazardous waste sites in several OECD countries indicates what environmental neglect could mean for other countries. While many of the environmentally friendly technologies and practices developed by OCED countries could be adapted to the needs of developing countries such as Sri Lanka and allow them to side-step environmentally detrimental development paths, OECD countries will have to assist, both in financial and educational terms, in the transfer of these environmentally sound technologies.

Note

1. World Commission on Environment and Development (WCED) *Our Common Future* (Oxford: Oxford University Press, 1987).

15 Trade and the Environment: A Sri Lankan Perspective

V. Kanesalingam

The significant expansion of trade and growth of tourism in Sri Lanka over the past decade has led to increasing concern about the impact this is having on the environment. The rapid expansion of industrial production has been emphasised without corresponding attention being paid to environmental protection. Moreover the primary concern of foreign investors and their local collaborators has been the promotion of exports, rather than ensuring environmentally sound industrial development. Tourism has also contributed to, and has been affected by, environmental degradation. This chapter will focus on these issues, with particular emphasis on their impact in Sri Lanka.

FREE TRADE AND THE ENVIRONMENT

15 December 1993 will be remembered as the date of the culmination of the greatest round of international trade negotiations and the date that brought to life the World Trade Organisation. The key objective of these multilateral negotiations was to establish a framework that would dismantle the remaining tariff and non-tariff barriers to international trade.

However some remain sceptical about whether the establishment of this new international organisation and the creation of a more comprehensive set of international trading rules will open the way for free international trade, or whether it will cement the existing global power structure. Indeed some believe that the new rules will restrain the control that states have over their own affairs and at the same time allow transnational corporations to become modern-day dinosaurs, able to stalk the new 'free trade' world and forage for opportunities wherever they can be found.

But with the conclusion of the Uruguay Round and the springing to life of the World Trade Organisation, smaller states dependent on trade can no longer choose managed economic systems over liberalised international trade. It is now a world of increasing competition and market accessibility. Comparative advantage lasts only as long as it takes for other trading nations to fill the breach. If there can be no turning back from liberalised trade, developing nations must find ways to encourage their industries to compete on the free trade court. First, there is a need for diversification of products and markets by all firms, be they local or transnational corporations. The first casualties of free trade will be those industries that have not outgrown protection and achieved efficiency and maturity. With the elimination of local or small regionally fragmented markets and the creation of a global free trade zone, fewer opportunities will exist for disguising inefficiency or protecting home-based industries. Although the power that can be wielded by transnational corporations may be problematic, there does not appear to be anything intrinsically preferable in a home-based enterprise. The comparison is usually based on their scale of operation, but both types of enterprise are commercially orientated.

Problems also arise from those who have an incomplete understanding of the implications of harnessing comparative advantages to make gains in international trade. Some economists appear to hold the view that nations that decide to 'pollute now and clean up later' are wisely using their environment as a comparative advantage. However a comparative advantage that is based on degradation of one's environmental assets can only be short-lived. Nations and their leaders must take care to appreciate fully the nature of their assets to ensure that they are not selling themselves short. This has happened in the past and we have a responsibility to ensure that it does not happen again. Continued prosperity will only come from comparative advantage based on knowledge, skills, labour-intensive technology and the efficient adding of value to wisely managed natural resources. Consequently, appropriate environmental regulations and safeguards must form part of any country's strategy for growth.

In addition the international market place, represented by consumers in all countries, is becoming increasingly concerned about the environmental degradation that is caused by the manufacture of the final product. Products and production processes that create pollution and endanger biodiversity anywhere in the world may not be acceptable to the environmentally conscious consumer. An integrated global market will seek to create and impose standards that will necessitate the judicious use of resources. If a country does not take the opportunity to modernise its technological capability so as to address the ecological concerns of consumers, it may once again miss the boat to prosperity.

Apart from the conclusion of the Uruguay Round and creation of the World Trade Organisation, the early 1990s also witnessed two other significant international trade developments: the establishment of the European Union (EU), comprising twelve very powerful countries in Western Europe, and the conclusion of the North American Free Trade Agreement (NAFTA), comprising Canada, the United States and Mexico. Although the EU and NAFTA are regional trading blocs that seek to promote free trade within their own respective geographical boundaries, there will no doubt be a significant impact on the trade links that the constituent member states have with the rest of the world.

These trade agreements, as well as the Uruguay Round of GATT, have not been without their environmental critics. For example full-page newspaper advertisements ran in the US press crying 'Sabotage of America's health, food safety and environmental laws'. Environmental organisations were concerned that the Uruguay Round agreements and NAFTA would reduce the effectiveness of health and food safety laws in the United States. Examples of these laws include those that severely limit the amount of pesticides permitted in food products and others that do not permit the addition of any cancer-causing agent to processed food. However the United States is not the only country where environmentalists are concerned about a relaxation of food safety standards. The Sustainable Agriculture, Food and Environment Alliance and the Public Health Alliance in the United Kingdom published a document in 1992 that expressed some of the same concerns articulated by United States environmentalists.

TRADE AND THE ENVIRONMENT: SOME VIEWS FROM SRI LANKA

There is no denying the fact that there are both costs and benefits from trade liberalisation and trade expansion. From an environmental perspective, trade liberalisation can certainly have an adverse effect on the environment. Sri Lanka's own development policies and trade strategies, as in most developing countries, seem to have failed to provide an adequate safeguard for the environment. There has been little realisation that long-term and sustainable free trade policies cannot be at the expense of the environment.

The 'greens' in Sri Lanka generally share the objections of environmentalists abroad about the environmental consequences of trade liberalisation and the Uruguay Round agreements. They are particularly concerned that international trade rules:

1. will limit national sovereignty and by doing so also limit the right of countries to apply whatever environmental measures they choose;
2. do not allow countries to ban the import of a product because of the way it is produced or harvested;
3. will prevent a country from imposing countervailing duties on imports produced under lower environmental standards than their own, and also discourage subsidies, which are one way of compensating producers for meeting higher environmental standards than their rivals;
4. will encourage the harmonisation of product standards which will impose higher standards with regard to food additives or pesticide residues and this could act as a trade barrier;
5. will prevent countries from imposing export bans, which they may want to use to protect, say, their forests or elephants. American environmentalists, for example, want to ban the exportation of certain pesticides that are prohibited in the United States but sold to developing countries;
6. will frown upon the use of trade measures to influence environmental policy outside a country's territory. Yet increasingly the issues that arouse environmental passion are those affecting 'global commons' – the oceans and atmosphere, animal and plant species threatened with extinction – that concern all countries;
7. may undermine international environmental agreements through their prohibition of trade measures that discriminate against individual nations. Yet such measures may be the most effective way for countries that play by the rules of an international agreement to penalise others that do not. Environmentalists want a set of carefully defined circumstances in which trade measures will be permitted under international environmental agreements;
8. will resolve disputes in a secretive way, without allowing environmentalists to put their arguments forward and without making important papers and other information available to them.

Arthur Dunkel, former secretary-general of GATT, accepted the need for international trade rules to be 'greened', although he first wanted to complete the Uruguay Round. Since the completion of the Round there has been a flurry of environmental activity within WTO. But WTO is no stronger than the consensus among its members. On the integration of trade and environmental policies and resolution of the issues noted above, there is as yet no consensus either among WTO member countries or within them.

CASE STUDIES OF RAPID INDUSTRIALISATION IN SRI LANKA

In addition to environmentalists' general concern for the effect of international trading rules on the environment, there are also more specific concerns relating to the effect of international trade on Sri Lanka's environment. Several illustrative cases can be cited.

The first case study is one where a large consignment of milk powder imported by a multinational firm was found to be contaminated by radioactive materials at a level much higher than that permitted by international standards. The milk powder had been imported at a low price from a non-conventional source, but was finally ordered to be reshipped back to the exporter. It was clear that although liberalisation of Sri Lankan trade rules had made it possible for goods to be purchased from non-traditional sources at cheaper prices, liberalisation also required more intensive scrutiny of the goods being imported to ensure that they met national and international health standards.

There have also been increasing complaints from environmentalists that many industries that are part of the country's programme of export-led growth do not comply with Sri Lankan environmental laws. It appears that the authorities have deliberately waived both environmental and labour standards, especially for those industries operating in certain free trade zones or export promotion zones. These industries have not been required to undertake environmental impact assessments relating to their location, measures for disposal of waste, air and noise pollution control and protective measures for their workers' health. In one example a testimony was presented at the Permanent People's Tribunal in Italy about a Sri Lankan fashion knitwear company that had exposed its workers to chemicals and fumes in the dyeing section, where they were not equipped with safety gear and prescribed safety measures were not being followed. According to the testimony, the managers of the company had also ignored workers' complaints about skin diseases and respiratory problems. This example of a company exposing its workers to hazardous chemicals and failure on the part of the authorities to enforce safety and environmental standards in export processing zones is not an isolated one, but is common in many Sri Lankan industrial units producing for export. Indeed the Ratmalana-Moratuwa suburb on the southern periphery of Colombo is plunging deeper into an environmental quagmire, with authorities and industrialists turning a blind eye to the health hazards caused by toxic factory effluents. Stronger environmental laws appear to be required, along with more vigilant action and enforcement on the part of the Central Environment Agency.

Environmentalists have made strong objections to the proposed Colombo–Katunayake express highway, which will connect the airport to Colombo. The government has determined that there is a need for a new highway to provide a quicker and more dependable corridor for the growing number of tourists and volume of freight between the airport and Colombo city. Environmentalists, however, are concerned that construction will involve demolition of around 1500 homes, destruction of thousands of coconut trees and other food-producing trees and general degradation of the environment. Around 2500 families living directly in the path of this 31 kilometre expressway will be dislocated. Those who live alongside the expressway will be cut off from neighbours, friends, relations or children who live on the other side, as there will be only a few pedestrian crossing points along the whole 31 km stretch. They will also be deprived of direct access to paddy fields, coconut palms and other properties on the other side. A 1992 environmental impact assessment for the project revealed that during the construction phase, land and water would be contaminated by waste while drainage paths and the ponding of water would also create severe health hazards. Once constructed, the expressway will be intolerably noisy and dusty for those living close by. It will also form a physical barrier to the flood waters that often pass through the region and could increase the region's susceptibility to severe flooding. This case is a clear example of the effect that increased trade can have on the environment.

Tourism is an important industry in Sri Lanka and is one of the more significant earners of foreign exchange. Tourist numbers have almost doubled over the past decade and continue to rise, and foreign exchange earnings having risen tenfold. There is, however, a two-way interaction between tourism and the environment. An attractive environment in places of tourist interest will bring tourists, but they will be repelled by the lack of sanitation, filthy roads and poorly maintained hotels, tourist houses, beaches and parks. However tourists themselves can cause serious degradation of the environment. Moreover tourist operators need to be aware of the impact that changes in the environment may have on tourism. For example increased ozone depletion and climate change may lead to a decline in tourism, as will beaches that are polluted by oil slicks or tar balls released from passing tankers.

Although Sri Lanka has passed a number of laws intended to reduce further degradation of the environment, such as the Environmental Impact Assessment Act, tourism development laws and pollution control laws, more needs to be done, especially with regard to monitoring and enforcement.

CONCLUSIONS

Free trade cannot be halted or reversed. But a process of unrestricted free trade and inadequate policy measures to control environmental degradation will only lead to disaster. The examples cited above do not support a case for restricting import–export trade or tourism. Rather they stress the need for strong policy prescriptions to ensure that industrial development and tourism are environmentally friendly. In this way trade and environmental protection can be made mutually compatible, and ecologically compatible industrial and tourist activities can provide secure investment opportunities for the future.

Part III

Paths Forward

16 Trade, Environment and the Transfer of Environmentally Sound Technology

Veena Jha and Ana Paola Teixeira*

It is generally understood that preservation of global and local environments will require the internalisation of currently external environmental costs. This need has manifested itself in both multilateral environmental agreements (MEAs) and in the formulation of environmentally orientated product and processing standards. Such standards are often devised to encourage producers to use environmentally sound technology (EST) or change their product or processing designs so that they are more environmentally friendly.

Differences between nations concerning their need to internalise these environmental costs has led to diverging standards. Concern in some countries about the level of standards in others have led to the use of trade measures in an effort both to improve standards and to offset the difference in compliance costs borne by domestic producers. Whether the use of trade measures in these cases is legitimate and consistent with current international trading rules remains a controversial issue. However, while this controversy remains unresolved, the market access of foreign producers, particularly those from developing countries, continues to be seriously threatened by new requirements to comply with these standards.

In the case of developing countries, part of the solution to this problem may lie in the development and transfer of EST, and in the setting of higher domestic environmental standards. This approach could enhance trade and at the same time help preserve the environment by improving the international competitiveness of developing countries, guarding local environments in these countries, and by having an environmentally neutral effect on the global commons. In contrast, limited access to EST could exacerbate the injurious effects of the use of trade measures against developing countries.

Unfortunately, at present there seem to be several barriers obstructing access by developing countries to EST. In the first place, as no distinction is made between EST and ordinary technology, there is concern that past reluctance to transfer ordinary technology will mean the same for EST. In addition, while particular ESTs may operate successfully in OECD countries, it appears that their selection and application in developing countries will have to be handled very carefully if they are to have the same success. There is also concern about the viability of ESTs if they are being applied only in the export sector. Finally, it appears that access to EST will be difficult because of cost, patenting and licensing controls, and to a lesser extent the way standards are formulated in some OECD countries.

THE URGE TO PROTECT GLOBAL AND LOCAL ENVIRONMENTS

A distinction must be made between global and local environmental concerns. While the former may necessitate major international initiatives, national or local operations are often sufficient for dealing with the latter. However this may not always be the case. Sometimes global problems have diverse regional environmental impacts that require different localised action. In addition, actions once thought to have only local effects may cause global problems over time and problems initially thought to be serious may come to be seen as less so.[1] However one thing appears clear: irrespective of whether environmental problems are local, regional or global in nature or effect, there is a strong desire on the part of all nations to address the environmental crisis.

Protecting the global commons

Responsibility for protecting the global commons is not clearly assigned at the international level. One reason for this is that it is not entirely clear whether user rights or property rights define ownership of the global commons. In the case of the ozone layer, both user and property rights can be said to be global. But in other cases, such as tropical rainforests, while the user rights may be global, ownership is clearly vested with either national governments or local communities. Therefore, when trying to resolve environmental problems in the global commons, issues of territoriality and national sovereignty become important.

The need to solve global environmental problems, though this may infringe on national sovereignty, has led to the conclusion of a number of multilateral environmental agreements (MEAs). Sometimes MEAs include trade provisions to assist in their enforcement and encourage the accession of nations that have not joined the agreement, thus preventing these 'free riders' from taking advantage of higher domestic environmental standards in foreign countries. A clear example of this is found in the Montreal Protocol on Substances that Deplete the Ozone Layer, which restricts trade in ozone-depleting substances between signatory and non-signatory nations.

While it is possible to argue that trade restrictions are less efficient than instruments that correct environmental distortions directly,[2] the perceived ease of use of these measures makes them a preferred instrument for the enforcement of MEAs.[3] Many developing countries and smaller countries are concerned that the use, or even the threatened use, of these measures will detrimentally affect their market access.

While some developing countries have expressed concern that in signing MEAs they could limit their development options, or be faced with additional financial burdens, the drafters of these agreements have on occasion sweetened the medicine by including provisions for the transfer of funds and technologies that have offered advantages to signatory developing countries. A case in point is the experience of China, which not only met but exceeded the targets set by the Montreal Protocol. Chinese firms used finances from the Multilateral Fund set up under the Protocol to manufacture and export low-cost refrigerators and air conditioners that did not use freon.

Protecting local environments

The protection of local environments is generally regarded as a domestic concern for which countries must set their own national environmental priorities. These policies are increasingly taking the form of product and process standards that apply to domestic and foreign producers and their products alike. Again, many developing countries and smaller countries are concerned that the use, or even the threatened use, of measures taken to uphold national environmental standards will detrimentally affect their market access.

For example, in 1991 US Senator Boren introduced the International Pollution Deterrence Act, which considered inadequate pollution controls, including inadequate enforcement of such controls, as subsidies. Countervailing duties equal to the amount it would cost a foreign firm to comply with US environmental standards were to be imposed on imported products. Other US proposals included the Global Clean Water Incentives Act, which would have required the secretary of commerce to impose fees on imported products subject to or manufactured from processes that did not comply with the US Clean Water Act standards and House Majority Leader Richard Gephardt's 'Green and Blue Super 301' proposals.[4]

If legislation such as this were to be enacted in the United States or in other OECD countries, then developing countries exporting to these countries would be obliged to change their product designs and processes to conform with the new legislative requirements. One consequence would be the immediate need to import environmentally sound technology.[5]

ENVIRONMENTALLY SOUND TECHNOLOGY

While the concept of an environmentally sound technology appears fine at the strategic level, the question of whether any particular technology is 'environmentally sound' at the operational level is very complex.

Because the operation of any technology is a mix of hardware components such as machinery or equipment, and software components such as knowledge and managerial practices, the combination of any hardware component may have different effects depending on the complementary software component and the conditions under which they are operated. Consequently the environmental soundness of any particular technology will have as much to do with the way it is used or where it is used, as with any inherent characteristics.[6] Moreover any specific item of environmental technology may have a superior performance for some purposes, while being less suitable for others.

In addition, sounder environmental technology may emerge over time and supplant existing technology. Since evidence of the environmental soundness of a particular technology is limited by prevailing scientific knowledge, the long-term effects or increased scale of use of any particular technology may cause unforeseen adverse environmental effects. A case in point are CFCs. These substances were initially used as substitutes for toxic substances such as ammonia and sulphur dioxide. For many years CFCs were considered safe and it was only much later, after their use had increased, that their damaging impact on the ozone layer was discovered.

A further question relates to the degree of environmental soundness. For example a broad definition may demand technologies that promote economic growth, create employment, are resource and energy efficient and reduce the generation of waste.[7]

Environmentally sound technology may be classified as 'end-of-pipe' or clean production technology. 'End-of-pipe' technology consists of equipment that is installed in existing machinery to remove pollutants from waste streams at the end of the pipe, following a 'clean-up afterwards' approach, while clean production technology involves changes in the production process and product composition in order to avoid pollution in the production process, following a 'pollution prevention approach'.[8]

Finally, the choice of technology or process may be affected by local environmental regulations – sometimes detrimentally. For example Celanese Mexicana inaugurated a new industrial complex in the 1980s with eight different chemical plants. The Mexican authorities insisted that an elaborate system be used to reduce air-polluting emissions from the plants, despite complaints form the US parent company. The resulting production process turned out to be less efficient and more polluting than similar plants in the United States.[9]

FACTORS INFLUENCING THE MARKET FOR EST IN OECD COUNTRIES

The market for EST in OECD countries has been driven primarily by legislative pollution controls, with more stringent controls being imposed over time. Public and consumer concern in OECD countries has also encouraged industries to adopt more environmentally responsible practices, thus safeguarding their commercial credibility and market share. Recently corporate image and the benefits of environmental investment have also become more important.

Technological innovation and markets for EST have tended to develop in countries with comprehensive environmental legislation. Demand for various kinds of environmental equipment has normally reflected national priorities and national industrial structures, with differences in standards leading to variations in the design and focus of ESTs. For example German expertise in water-treatment equipment can in part be attributed to stringent national regulations dealing with water pollution control.[10]

The main purchasers of environmental equipment and services in OECD countries have been municipalities, power and water utilities, the mining industry and several traditional manufacturing sectors. Expenditure is marginally higher on average in the public sector than in the private sector,[11] with the public sector allocating a higher share of its expenditure to water treatment and waste disposal compared with the private sector, whose purchases are mainly in air pollution control.[12] In the private sector the share of investment expenditure on pollution control has been estimated at between 2 per cent and 4 per cent of total manufacturing investment, although this figure appears considerably lower than it was in the 1970s, when environmental legislation was first enacted.

The market growth for the OECD environment industry in non-OECD countries is projected to be somewhat higher during the 1990s than for the OECD area itself, due to a rapid increase in demand in the Asian and East European markets. For example Hong Kong, Taiwan and South Korea are planning to spend around US$5 billion over the next five years. The World Bank and the International Finance Corporation are also funding more environmental projects and now require environmental impact assessments for industrial projects in developing countries, further stimulating demand for EST. Moreover firms supplying environmental goods and EST are benefiting from export credits and tied-aid financing.[13]

ENVIRONMENTAL STANDARDS IN OECD COUNTRIES

The demand by firms in developing countries for environmentally sound technology appears to be mainly due to their need to develop products and processes that comply with environmental standards or consumer preferences in their major markets. Thus the way in which these standards are set and consumer preferences are manipulated is of vital concern to developing countries.

If standards in OECD countries are set domestically, the main actors involved in the standard-setting process are usually central and local government, and industry and citizen groups, although the influence of each of these actors on the standards being set differs from case to case.

Although the nature of the interaction between government and industry in standard setting differs between countries and regions, industry often has a significant role in the determination of environmental standards.[14] Governments often involve industry very closely in the determination of standards,[15] and in some cases rely on industry's knowledge and technical competence when developing standards.[16] The role of industry appears even more dominant in cases where there are divergent views about the significance of various ethical, ecological and economic variables, and where there is scientific uncertainty about the outcome of any particular environmental problem.[17]

This process of standard determination has important implications for developing countries. Above all, if standards are to be applied equally to all traded products, then the standards must be met by products exported from developing countries. Given that the setting of standards in OECD countries often involves industry, it follows that the standards will reflect the industrial practices of the country concerned and may often mandate the use of particular technology available only in that country.

This phenomenon has been clearly observed in the case of standard setting for eco-labelling. In general, criteria are set so that at least some existing domestic firms will have the technology that will enable them to obtain an eco-label. Thus firms from developing countries seeking to obtain a label for their products in that market may be forced to buy technology from firms already having the label.[18]

While this type of standard setting generally operates against competitive products from developing countries, the situation is often very different for products that are predominantly manufactured in developing countries alone. Here importers and retailers will have an interest in the standards being set so that imports will not be unnecessarily impaired. For instance this occurred in the case of leather products exported from India to Germany. During the late 1980s, Germany introduced an environmental standard that banned the use of PCP (pentachlorophenol) in leather tanning. Although it took over two years and an increase of about 20 per cent in capital costs to adjust to this new standard, because Germany was interested in supporting Indian leather exports and as tanning was rarely done in Western Europe, it was possible to implement and finance the necessary changes in India.[19]

While the development of standards under MEAs may arguably be more equitable because they are usually negotiated internationally, compliance with MEA standards may also require at least initial reliance on just a few technology suppliers from OECD countries.

CONSTRAINTS ON THE TRANSFER OF ENVIRONMENTALLY SOUND TECHNOLOGY TO DEVELOPING COUNTRIES

Although it may be essential for firms in developing countries to obtain environmentally sound technology, there appear to be a number of serious constraints on the transfer of this technology.

Purchase and licensing of EST

A large share of the international transfer of EST takes place through technology licensing rather than importing and exporting equipment. Many larger firms develop pollution control technology for their own use and then license the technology to other firms for production or internal use.

Generally firms in developing countries obtain access to new technology by licensing or assembling the technology from a blueprint.

The single most important factor inhibiting the transfer of EST to developing countries is its cost, exacerbated by difficulties in obtaining bank credit and foreign exchange. Poor cash flow may also prevent firms from purchasing pollution prevention technology, even though it may offer lower operating costs and improved productivity over the long term.[20]

The market for ESTs is quite small, leading to further costs in the transfer process. A small number of large firms account for about 50 per cent of the market, with a large number of smaller firms accounting for the remainder.[21] Given that the market for EST is already supply driven, firms from developing countries may find it difficult to locate the smaller firms, largely because of a lack of information about these firms and their technology. Thus limited access to the suppliers of EST may lead to a situation where an already cartelised market yields even higher monopoly rents for the suppliers. Under these circumstances it is possible that the cost of obtaining EST may not reflect the true opportunity costs of EST alone.

Apart from the direct cost of acquiring EST, there may also be several indirect costs. These include the need for additional skills, infrastructural facilities and/or management practices that may be required to operate the technology. Furthermore EST that responds to environmental standards based on life-cycle analysis may require firms to undertake backward integration. In this case large outlays of capital will be required. Moreover firms may have to use one type of technology for domestic production and another for export production, reducing the benefits of economies of scale. Finally, obtaining information about environmental standards and available EST may also be expensive.

It has been shown in OECD countries that the capital costs of EST are generally low and do not affect international competitiveness. While the costs in OECD countries may be relatively small, they are certainly higher for certain industries, such as copper smelting and refining. Indications are that the installation and operation of pollution control and abatement equipment for these industries may have a negative impact on competitiveness.[22]

In an OECD survey of seven industries, it was found that the most important obstacles to the transfer of clean technology to developing countries were cost and the lack of environmental regulations.[23] The survey also found that in several cases clean technology offered no cost advantages over existing or traditional production technology, leaving firms in developing countries little incentive to import them.

While the operating costs of EST in certain industries, such as leather tanning, textiles, toys and furniture, may not be very high, capital costs may range from 15 per cent to 25 per cent.[24] Thus firms in developing countries may tend to choose end-of-pipe technology, where capital costs are lower, rather than clean production technology, especially where there are large risks. This appears to have been the case with implementation of the Montreal Protocol. Over 60 per cent of the projects approved for funding under the Montreal Protocol involved retrofitting.[25] While this may not be the best solution for maintaining long-term competitiveness, it was an important economic opportunity for firms in developing countries.

Finally, concern must also be raised about the low level of transfer of ordinary industrial technology to developing countries over the last decade, and whether the transfer of EST will fare any better.

Patents

Few ESTs appear to have been patented. Therefore patents have not been considered a serious barrier to their transfer. Furthermore it appears that the ESTs that have not been patented are designed to address very specific problems and use very sophisticated technology that is not easily copied. The true cost of patents in the transfer of EST is not clear, nor is the effect of the costs of patents in the choice of technology. For example most transfers of EST under the Multilateral Fund of the Montreal Protocol have been of technologies where information about them is in the public domain, rather than those that involve patents.[26]

Foreign direct investment

There is an increasing tendency for EST suppliers to enter foreign markets through direct investment, cross-border mergers and acquisitions, joint ventures or collaboration with foreign partners. However this phenomenon is more strongly observed in OECD countries than in developing countries.

It appears that developing countries, with the exception of those in Asia, are becoming less attractive for foreign direct investment, including FDI for the transfer of EST.[27] One reason mentioned by investors is the lack of environmental regulations or the enforcing of existing regulations in developing countries.[28] In fact this seems to be encouraging foreign investors to consider transferring 'dirty' industries to developing countries.[29] For example there is evidence to suggest that some leather tanning processes have shifted to developing countries in South Asia in response to stronger environmental regulations on tanning chemicals and processes in Europe.

Further constraints on the transfer of EST through FDI result from the domestic structure of incentives in the host country. A number of foreign investors have complained that the prevailing structure of incentives is not enough for them to invest in the transfer of EST.[30]

Negotiations over specific transfer of EST may need to be accompanied by incentives such as tax holidays, grants and other forms of assistance on the part of host developing-country governments. For instance in developed countries host governments have often encouraged EST transfers by subsidising a percentage of the pollution control expenses, giving tax breaks on imported pollution technology or assisting in waste disposal.[31] For developing countries these types of incentive appear to increase the cost of acquiring EST.

CONCLUSIONS

A shortage of financial resources combined with increasing needs in developing countries could result in poor environmental choices being made. To preserve global and local environments, and to remain competitive in foreign markets, EST needs to be transferred to developing countries. However there is no agreement on the most appropriate mechanisms for transfer.

Clearly a number of constraints already stand in the way of the transfer of EST. This chapter has identified a number of difficulties, including those that arise from attempting to directly apply EST designed in OECD countries to the conditions that prevail in developing countries; barriers posed by standard setting in OECD countries that does not take sufficient account of the needs of exporting firms in developing countries; the cost of patents and of purchasing or licensing EST; and the difficulties associated with foreign direct investment

OECD countries prefer technology transfer on commercial terms with some financial assistance, while developing countries claim that the transfer of EST should be made on a preferential and concessional basis.[32] While there remains no consensus about the transfer of EST, a number of developed countries are directing the organisers of their aid programmes to consider this issue. Nevertheless the transfer of EST will also require participation from the private sector.

ESTs cannot be considered as just another set of technologies, but are a special group with distinct, inherent characteristics. Their transfer from developed to developing countries will be an essential requirement of sustainable development.

Notes and references

* The authors work for the International Trade Division and the Division for Science and Technology of UNCTAD. The views expressed in this paper do not necessarily reflect those of UNCTAD. The authors wish to acknowledge the helpful comments of H. V. Singh and Grant Hewison.

1. See Box 3-A, 'The Global-Local Continuum', in Office of Technology Assessment. *Trade and Environment: Conflicts and Opportunities*, US Congress, OTA-BP-ITE-94 (Washington, DC: OTA, 1992). See also D. Robertson, 'Trade and Environment: Harmonization of Technical Standards', in World Bank, *International Trade and Environment*, P. Low (ed.) (Washington, DC: World Bank, 1992).
2. J. Bhagwati and T. N. Srinivasan, *Lectures on International Trade* (Cambridge, Mass: MIT Press, 1983).
3. OECD, 'The OECD Environment Industry: Situation, Prospects, and Government Policies', OCDE/GD(92)1 (Paris: OECD, 1992).
4. Office of Technology Assessment, *Trade and Environment: Conflicts and Opportunities*, U.S. Congress, OTA-BP-ITE-94 (Washington, DC: OTA, 1992), p. 92.
5. It has also been argued that there is no reason to assume that the transfer of EST must necessarily take place from the North to the South. Avenues for the promotion of the environmental virtues of some of the existing traditional technologies (mostly in developing countries), particularly with respect to agricultural products that are growth inducing, should be explored. See Kojo Amanor, 'Indigenous Technology and Natural Resource Management Systems', paper presented at an UNRISD Conference on Social Dimensions of Environment and Sustainable Development, Malta, 1992.
6. See A. Rath and B. Herbert-Copley, 'Green Technologies for Development: Transfer, Trade and Cooperation', Searching Series 6 (Ottawa: IDRC, 1993); and C. Almeida, 'Development and Transfer of Environmentally Sound Technologies in Manufacturing: A Survey', Discussion Paper No. 58 (Geneva: UNCTAD, 1993).
7. For a comparison of the definitions of ESTs adopted by UNCED and the OECD, see A. Barnett, 'The International Transfer of Technology and Environmentally Sustainable Development', paper prepared for the UNCTAD Workshop on the Transfer and Development of Environmentally Sound Technologies, Oslo, October 1993.
8. C. Almeida, 'Development and Transfer of Environmentally Sound Technologies in Manufacturing: A Survey', Discussion Paper no. 58 (Geneva: UNCTAD, 1993).
9. H. J. Leonhard, *Pollution and the Struggle for the World Product – Multinational Corporations, Environment and International Comparative Advantage* (Cambridge: Cambridge University Press, 1988).

10. OECD, 'The OECD Environment Industry: Situation, Prospects, and Government Policies', OCDE/GD(92)1 (Paris: OECD, 1992).
11. Ibid.
12. Ibid.
13. US Congress, Office of Technology Assessment, *Trade and Environment: Conflicts and Opportunities*, OTA-BP-ITE-94 (Washington, DC: OTA, 1992).
14. Within the OECD, the differences are the largest between North America, Western Europe and Japan. However within Europe the differences are striking between the Netherlands and the Scandinavian countries – where the political culture is characterised by the pursuit of consensus between government and business groups – and Germany, France and Britain. For a more complete explanation, see G. Kleeper, 'The Political Economy of Trade and Environment in Western Europe', and c. Van Grasstek, 'The Political Economy of Trade and the Environment in the United States', in World Bank, *International Trade and Environment*, P. Low (ed.) (Washington DC.: World Bank, 1992).
15. M. Iba, 'Japanese Environmental Policies and Trade Policies: Trade Opportunities for Developing Countries', paper prepared for UNCTAD/UNDP project, 1992.
16. 'To gather information and to obtain cooperation instead of obstruction, the government has hardly any choice but to go for a consultation with the industry involved' argues Verbruggen in H. Verbruggen, 'The Trade Effects of Economic Instruments', paper prepared for the OECD Environment Directorate/NACEPT Trade and Environment Committee, Informal Experts Workshop on Environmental Policies and Industry Competitiveness, Paris, 28–29 January 1993.
17. O. Kuik and H. Verbruggen (eds), *In Search of Indicators of Sustainable Development* (Dordrecht and Boston: Kluwer Academic Publishers, 1991).
18. V. Jha, R. Vossenaar and S. Zarrilli, 'Eco-labelling and International Trade: Preliminary Reports from Seven Systems', paper prepared for the ISO sub-group on Labelling, London, May 1993.
19. I. Scholz and J. Wiemann, *Ecological Requirements to be Satisfied by Consumer Goods – A New Challenge for Developing Countries' Exports to Germany* (Berlin: German Development Institute, 1993).
20. US Department of Commerce, 'A Competitive Assessment of the U.S. Industrial Air Pollution Control Equipment Industry', (Washington DC: International Trade Administration, August 1990), Table 20.
21. OECD, 'The OECD Environment Industry: Situation, Prospects, and Government Policies', OCDE/GD(92)1 (Paris: OECD, 1992).
22. OECD, *Macroeconomics Evaluation of Environmental Programmes* (Paris: OECD, 1978).
23. OECD, 'Trade Issues in the Transfer of Clean Technologies', Technology and Environment Programme, OCDE/GD(92)93 (Paris: OECD, 1992).
24. I. Scholz and J. Wiemann, *Ecological Requirements to be Satisfied by Consumer Goods – A New Challenge for Developing Countries' Exports to Germany* (Berlin: German Development Institute, 1993).
25. See *Multilateral Fund for the Implementation of the Montreal Protocol – Inventory of the Approved Projects* (Montreal: June 1993).

26. O. E. El-Arini, 'Technology Transfer and the Montreal Protocol', paper presented at the Conference on the Montreal Protocol for the Indian Industry, New Delhi, India, 28–29 January 1993.
27. UNCTAD, *World Investment Report 1993* (New York: United Nations, 1993).
28. OECD, 'Trade Issues in the Transfer of Clean Technologies', Technology and Environment Programme, OCDE/GD(92)93 (Paris: OECD, 1992); and Congressional Research Service, *Financing New International Environmental Commitments*, report prepared for the Committee on Foreign Affairs of the US House of Representatives and the Committee of Foreign Relations of the US Senate by the Congressional Research Service, Joint Committee Print, (Washington, DC: US Government Printing Office, 1992).
29. World Bank, *International Trade and Environment*, P. Low (ed.) (Washington DC: World Bank, 1992).
30. H. J. Leonhard, *Pollution and the Struggle for the World Product – Multinational Corporations, Environmental and International Comparative Advantage* (Cambridge: Cambridge University Press, 1988).
31. Industrial Development Authority (IDA), *Industrial Plan 1978–82* (Washington DC: IDA, 1988).
32. *Agenda 21*, A/CONF.151/PC/100/add.9, Fourth Session of the Preparatory Committee for the UNCED, 2 March to 3 April 1992, Section 1V, Chapter 2.

17 The Transfer of Environmentally Sound Technology with Special Reference to India

Amrita N. Achanta, Pradeep Dadhich, Prodipto Ghosh and Ligia Noronha*

Although there is increasing acknowledgement that the development and transfer of environmentally sound technology will be an essential component of the strategy to achieve sustainable development, little is actually known about how to promote and facilitate such development and transfer. It does seem clear, however, that if the failed policies used to 'facilitate' the transfer of conventional technology is maintained and applied to the transfer of environmentally sound technology, there is little reason to believe that any significant or rewarding transfers will take place.

Some serious questions need to be answered. In particular, can we learn any lessons from the transfer of conventional technology that may assist us in developing better policies for the transfer of environmentally sound technology? Indeed what is environmentally sound technology and how does it differ from conventional technology? Moreover, what are the forces that are creating the need for developing-country firms to adopt environmentally sound technology? And finally, what types of policy and policy instruments do we need to facilitate the transfer of these new types of technology? This chapter will examine these questions using a case study approach and selected Indian export firms.

THE TRANSFER OF TECHNOLOGY: SOME GENERAL ISSUES

Prior to moving to a discussion of environmentally sound technology it is important first to examine some of the issues that relate to the transfer of technology generally. Thus, this part of the chapter will examine the mechanisms by which technology is usually transferred, the nature and extent of technology transfer examined from a global perspective, and some of the past barriers to the effective transfer of technology to Indian industries.

The transfer of technology is generally understood to be the process by which technology or knowledge developed in one organisation is applied and utilised in another. Bell distinguishes three categories of transferable technology: (1) capital goods, services and other design specifications; (2) skills and know-how for production; and (3) knowledge and expertise that generate technical change.[1] There are several mechanisms, both commercial and non-commercial, by which these three categories of technology can be transferred from a supplier to a recipient. The principal routes of commercial transfer include foreign direct investment, joint ventures, licensing of the property rights to the technology, outright purchase, technical assistance, sale and subsequent servicing of the technology and franchising of the goods and services that can be produced with the technology.

While it is difficult to be precise about the nature and extent of global technology transfer, an approximation can be obtained by examining the extent of commercial flows. Almeida has estimated that in 1988 the global exchange of capital goods, foreign direct investment, technology payments, and technical cooperation grants amounted to US$700 billion, US$161 billion, US$48 billion and US$13 billion respectively. Of this, the developing country share was 21 per cent of the capital goods import, 18 per cent of foreign direct investment and 97 per cent of the technical cooperation grants.[2] However it is very difficult to estimate the quantum of these transfers undertaken to comply with environmental standards or regulations.

The success of any particular transfer will depend on the conditions present within the recipient firm's country as well as the firm. These often include factors such as the stage of technical development of the firm and the country, the characteristics of the users of the technology and the potential for absorption and diffusion of the technology into the country concerned.

THE TRANSFER OF ENVIRONMENTALLY SOUND TECHNOLOGY: SOME KEY ISSUES

Ordinarily the expression 'environmentally sound technology' brings to mind some kind of technical equipment that will be attached to other production equipment in order to bring about a more sound environmental outcome. However the concept of environmentally sound technology is in fact much wider than that and has been defined by Agenda 21 of the Earth Summit as total systems incorporating know-how, procedures, goods and services, and equipment, apart from organisational and managerial procedures. The Business Council for Sustainable Development, for instance, has identified good housekeeping, materials substitution, manufacturing modification and resource recovery as four prevention strategies aimed at a better environment.

The UNCTAD *ad hoc* working group on the Interrelationship between Investment and Technology Transfer has identified the transfer of environmentally sound technology as one of the key issues of UNCTAD's work programme.[3] The working group directed that further study of the transfer of environmentally sound technology be urgently undertaken with a focus on identifying the issues involved in the interlinkages between environment and technology; examining effective ways in which environmentally sound technologies could be accessed and transferred, particularly to developing countries; the possibilities for financial schemes and mechanisms that would assist with the transfer, development and improvement of these technologies; analysing the implications of environmentally sound technology for international competitiveness; and promotion of interchange of experiences among countries, apart from encouragement of initiatives aimed at sustainable development.

As mentioned previously, this chapter will discuss some of these issues by examining the transfer of environmentally sound technology to India, the earlier barriers that restricted transfer and the policies being developed to overcome these barriers.

FORCES OUTSIDE INDIA OPERATING TO ENCOURAGE THE TRANSFER OF EST

The following section of this chapter examines forces and policies outside India that are encouraging the transfer of environmentally sound technology. Although domestic forces also encourage this transfer, they will not be specifically addressed.

Many of India's export destinations are adopting environmental policy frameworks requiring Indian export firms to adapt their production processes to suit the environmental concerns of those markets. This study will, however, focus only on the European Union and Japanese markets.

The environmental framework of the European Union

The European Union's Environment Policy was adopted in 1972. The objectives of the policy include the prevention, reduction and as far as possible elimination of all pollution and nuisances; ensuring the sound management of and avoiding any exploitation of resources of a nature that may cause significant damage to the environment; ensuring that more account is taken of the environmental aspects of town planning and land use; and seeking common solutions to environmental problems within states outside the EU, particularly through international organisations. The EU environmental policy includes more than 100 legislative acts relating to a wide range of environmental matters.[4]

One of the key features of EU environmental law is the attempt to harmonise environmental standards and technical regulations throughout the member nations. This is being achieved through the use of official regulations, directives, decisions and recommendations passed by the various EU organs, as well as setting standards and voluntary covenants within industrial sectors. This chapter confines itself to the relevant trade-related environmental standards and legislation.

EU environmental law can be very confusing for an exporter. Not only is it important to be aware of EU environmental legislation, but also the environmental laws of each member country, which in some cases may be more stringent.

The EU has now emerged as one of India's largest trading partners, with an approximate share of 25 per cent of India's exports. The main export items are textiles, yarn, fabrics, garments, jute, coir, gems, jewellery, leather and leather goods.[5] However in this chapter only textiles, leather, automobiles, chemicals, pharmaceuticals, chlorofluorocarbons and the packaging of goods will be briefly considered for their environmental factors.

Textiles

In the case of textiles and clothing, Germany is a major export market for India. Section 30 of the German Foodstuffs and Essential Commodities Act requires manufacturers of textiles or clothing to ensure that the only finishing agents used are those that will not make the product detrimental to health if they are used correctly. Further statutory measures in this area require that consumers be protected by compulsory labelling with regard to formaldehyde use that exceeds 1500 mg/kg, and a ban on the use of carcinogenic substances, asbestos yarns and PCPs. Apart from these examples, a number of other chemicals are prohibited for use in textiles in Europe as a whole.

Two eco-labels, MST and MUT, can be awarded for environmentally compatible textiles in Germany. MST sets standards for consumer goods and prescribes a lower content of dangerous substances in textiles based on the Oko-Tex Standard 100. The eco-label is awarded by the Association for Environment and Consumer Friendly Textiles. The MUT is awarded to intermediate textile products and sets standards for the production process.

One firm in India, Gokak Mills, located in Karnataka, has applied for Oko-Tex certification. Gokak Mills mainly exports dyed and grey yarn to Germany. Recently it obtained the technical knowledge from a German collaborator for dyeing cotton yarn in a more environmentally sound manner, and it now follows the processing parameters set by this German importing firm in order to comply with the German standards. Gokak Mills has also started to produce 'green' or 'organic' cotton that is grown without the use of synthetic pesticides or fertilisers. Certification for the export of 'green' cotton is currently being sought from the United States.

Leather

In both Germany and the EU new standards have set limits on the use of PCP in leather manufacture to 5 mg/kg and 1000 mg/kg respectively. There is also a limit on the use of formaldehyde in excess of 1500 mg/kg. Consequently firms exporting to Germany will need to be aware of the more stringent standard, while those exporting to other countries in Europe will be subject to the less stringent EU standard. Certain leather-producing firms in India have voluntarily undertaken not to use azo dyes.

Automobiles

The EU has adopted several measures concerning pollution caused by vehicle exhaust emissions and motorcycle noise that are applicable throughout the EU. The standards set levels for carbon monoxide and nitrous oxide emissions from petrol and diesel engines, and specific levels of these gases for cars below 1400 cc, as well as particulate emissions from diesel engines. There are also standards for noise pollution.

TELCO of India exports cars to the EU – mainly to France and Spain – and has a significant number of orders for Tatamobile pick-up trucks and Tata Sierra model cars. To comply with the EU vehicle emission and noise standards, not only did TELCO have to up-grade its existing technology and obtain certification, but also make sure that all of its component manufacturers and suppliers did the same. The major technological changes carried out were in the fuel pump, with minor modifications in the engine air intake line and adaptability of the engine to operate with the type of diesel used in EU countries. It appears in this case that most of the technological changes required were facilitated through information supplied by a technical consultant, with all changes being carried out by TELCO itself. TELCO is also planning to export to the United States, where it is compulsory for the exporter to meet US-EPA standards.

Chemicals

Both the EU and Germany have passed regulations relating to the use and packaging of dangerous substances.

The German Law for Protection from Dangerous Substances 1980 concerns substances that are poisonous, caustic, irritating, carcinogenic, or have qualities that may alter the characteristics of water, soil or air. These regulations also concern the packaging and labelling of such chemicals.

The EU has, through the European Council, also passed regulations that seek to harmonise member laws relating to the classification, packaging and labelling of dangerous substances. Council directive 92/32/EC requires chemical producers outside the EU to provide documentation, special packaging and labelling of dangerous substances. Article 22 of the directive states that dangerous substances cannot be marketed unless their packaging meets certain design and safety requirements. The labelling provisions of the directive require each package to be clearly and indelibly marked with the name of the substance and the name and address of the individual in the EU who has placed the package and product on the market.

A further Council regulation, no. 2455/92 of 23 July 1992, establishes a common system of notification and information requirements for imports from and exports to third countries of chemicals that are banned or severely restricted because of their effect on human health and the environment. The system applies the international notification and 'prior informed consent' procedures established by the United Nations Environment Programme and the Food and Agriculture Organisation. One chemical banned under this regulation is PCT (polychlorinated terphenyl) which is used in the Indian leather industry. Since the introduction of the regulation, supplies of this chemical from Europe have been affected.

Pharmaceuticals and enzymes

The pharmaceutical products exported by Karnataka Antibiotics and Pharmaceuticals (KAPL, which exports pharmaceuticals to Germany, Belgium, and Denmark) and Biocon India adhere to the US Pharmacopiea (22)/BP-93/USP/BP-93/BCP and the British Pharmacopiea, which lay down detailed specifications in terms of manufacturing processes, raw materials used and packaging aspects. These also specify methods of storage and special labelling requirements. The pharmaceutical products manufactured for export require process audits by the World Health Organisation (WHO) under the Good Manufacturing Practices (GMP) guidelines and under 21-CFR-10.90 (USA), which is set by the United States Department of Health and Human Services, the public health service of the United States Food and Drug Administration. The products have to be further evaluated by the US Food and Drug Administration, who register medical products for export. Biocon manufactures products that meet Joint Expert Committee on Food Additives (JECFA) specifications as well as the GMP.

In response to the new requirements in Europe, KAPL up-graded the technology used in its manufacturing process. The formulations for injections and tablets were modified to suit the BP/USP specifications. Minor alterations were made to the production process and sophisticated instrument analysis was introduced to control the quality of the formulations. In the case of tablets, raw materials were imported to meet purity requirements and special tamper-proof containers were also designed. Modifications were made to increase the product shelf life of some products from three to five years and a process audit certification was obtained from the World Health Organisation so that these products could be exported.

Chlorofluorocarbons

The EU's regulations or recommendations relating to chlorofluorocarbons (CFCs) include a European Council resolution of October 1988 that requires urgent action by EU members to limit the use of CFCs and halons in all products, equipment and processes. It recommended discussions with industry to substitute CFCs wherever feasible and also a voluntary agreement on a common EU label for CFC-free products. A recommendation of April 1989 by the Commission (89/349/EEC) also requires a reduction of CFC use by the aerosol industry. These EU actions are very important to those Indian exporters whose products use CFCs, and need to be adhered to by firms that are seeking eco-labelling to show that their products are CFC free.

Shriran Refrigerators is one of India's largest integrated air conditioning and refrigeration equipment manufacturers. So far most of the air conditioners use HCFC-22 as the refrigerant while all refrigeration compressors use CFC-12 as the refrigerant. It is proposed to replace the CFC-12-based compressors with compressors designed by Tecumseh, USA, which use HFC-134a as the refrigerant.

Subros, one of the largest manufacturers of automotive air conditioning systems in India, has a technical collaboration agreement with

Nippondenso of Japan, which has developed the technology to manufacture automotive air conditioners based on HFC-134a. This technology is being transferred to Subros.

Packaging of goods

While the EU has not yet itself dealt with packaging waste, several member countries have passed their own legislation, the most well known being the German Ordinance on the Avoidance of Packaging Waste, 1991,[6] although an amended proposal for a directive on packaging and packaging waste was submitted to the European Council in October 1993. This proposal seeks to harmonise national measures for the management of packaging and packaging waste. The directive establishes several targets and essential requirements that packaging must meet and applies to all packaging waste in the EU. It also takes into account both quantitative and qualitative aspects of packaging waste, as well as chemical composition, for instance the concentration of heavy metals in packaging and packaging waste must not exceed certain specified levels.

Before this new packaging directive came into force, packaging was dealt with under Council Directive 85/339/EEC of June 1985, which governed the production, marketing, use, recycling and refilling of containers of liquids for human consumption, and the disposal of used containers.

Gokak Mills has made several changes to both its material and packaging methodology so as to meet the German ordinance, including abandoning the use of polyethylene cone discs, hard board cone discs, polyethylene cone bags, polypropylene straps and metal seals, and switching over to pallet packing.

Outcomes of the case-studies

Some of the preliminary findings of the case-studies indicate that exporting firms are willing to upgrade their technology so that they may continue to export. Very often the upgrade has been carried out indigenously. On many occasions upgrading has also been required of suppliers. One firm has even gone to the extreme of importing drugs in bulk to meet the stringent purity levels required by the OECD. One requirement common to all firms is certification of their products prior to export. All the firms contacted in this study displayed a reluctance to reveal the exact nature of changes and the costs of the changes. Although one of the firms indicated that the quality of its exported product is identical to that produced for the domestic market, there is a definite need for Indian manufacturers to adopt more stringent environmental standards in future.

The environmental framework of Japan

Japan has emerged as a major market for some of India's traditional goods or raw materials, such as cotton and mineral ores. Japanese environmental law is much simpler than that of the EU, but nonetheless requires careful attention on the part of Indian exporters. For the purposes of this study, only the Foods Sanitation Law, Japanese agricultural product standards, Japanese industrial product standards, and the Japanese eco-mark will be briefly considered.

Foods Sanitation Law

In Japan, foods, additives, apparatus, container packages and certain toys are regulated under the Foods Sanitation Law of 1947, administered by the Ministry of Health and Welfare. The Foods Sanitation Law applies to imported foodstuffs as well as to domestic products. There are approximately 60 provisions relating to standards and testing procedures for imported foods, including requirements for labelling, standards for foods and additives as well as a ban on the addition of certain chemical additives. Unlike in the EU, the Foods Sanitation Law of Japan does not relate to the production process of any agriculture or fisheries product. Presently the law is undergoing review in light of several advances that have been made to the international standards contained in Codex Alimentarius.

Japanese agricultural product standards

Standards are generally set on a voluntary basis by the government or by product manufacturing associations in Japan. Japanese agricultural standards relate to a wide range of processed foodstuffs and forest products that are either manufactured in or imported into Japan. Quality standards are generally set for organoleptic criteria, such as appearance and properties; physical and chemical criteria, such as componts; net weight and volume; criteria concerning containers; and ingredients that are permitted in the product. There are also labelling standards. Pesticide residue standards are set for a number of agricultural products, and in 1993 standards were set for 74 pesticides covering 130 products. Some of these new standards will affect Indian cashew nut growers, who have a major export market in Japan.

Japanese industrial product standards

The Industrial Standardisation Law of 1949 provides the basis for Japanese industrial standards, which apply to nearly all industrial and mineral products. These standards are voluntary and cover imported items such as iron ore from India.

Japanese eco-mark

The Japanese eco-mark was established in February 1989 by the Japanese Environment Association, with guidance and assistance from the Japanese Environment Agency. The eco-mark, as with similar marks elsewhere, seeks to inform consumers about the environmentally friendly nature of a product. As of March 1993, 2450 products in 55 product categories were covered by the eco-mark. Eco-marks are available for goods produced with environmentally sound manufacturing methods, those which place a minimal burden on the environment, and those made from recycled material or recyclable materials. However there seems to be only a limited awareness of the eco-mark among Japanese consumers.

ISO Series/Standards

In addition to the standards being formulated at the national or regional economic level, Indian exporters also need to be aware of the standards being developed by the International Standards Organisation (ISO). Since 1979 the ISO has been developing a single standard for the operation and management of quality assurance and is beginning to build on this progress in the environmental field. ISO drafts 'systems standards' rather than 'product standards', which create only the framework by which quality control is undertaken within the production process, specifying developments at each stage. ISO standards do not fix standards for any particular product, but instead provide standards for quality control mechanisms. Although not yet mandatory, compliance with ISO standards is becoming increasingly important for exporters to the EU. The ISO 9000 Series does not directly deal with environmental protection, although environmental laboratories and consultancy institutions can obtain this certification. The ISO is presently preparing the ISO 14000 Series, which deals exclusively with environmental protection and the control of industry.

DEVELOPING AN INDIAN POLICY FRAMEWORK FOR THE TRANSFER OF ENVIRONMENTALLY SOUND TECHNOLOGY

As noted above, in just two of India's important export markets there are now significant environmental policies and standards that have current or potentially serious effects on the sale and competitiveness of Indian products. To comply with these environmental policies and standards, Indian firms will need to acquire new and up-to-date technology or technological procedures that lessen the adverse environmental impact of their production processes and the products they manufacture. While some of this technology may be found indigenously, others will need to be transferred to India from overseas. Questions arising relate to whether there are any trade barriers to the transfer of this technology and what governmental policies need to be in place to overcome these barriers and further facilitate technology transfer.

Barriers to the diffusion of technology

A number of barriers already exist in India to the diffusion of technology. These include the difficulties that arise from the very properties of the technology market itself, the complexity of the market for environmentally sound technology and financial constraints on Indian firms.

The inherent properties of the technology market are a significant barrier to buyers from developing countries. For instance they face difficulties in recognising opportunities, determining the value of the technology and developing the organisational capability to ensure absorption of the technology once it is purchased.[7] While the complexity of the market for environmentally sound technology and its recent development are significant barriers to the diffusion of such technology, the limited access to financial resources and investment by Indian firms may perhaps be a more significant impediment to the transfer of technology.[8]

Within Indian industry itself, several other barriers have arisen. These include the inadequate competitive pressures on local industry; a limited perception by local industry of the gains to be had from purchasing new technology; the non-inclusion of environmental costs in decision making about choices of technology, thus making EST more costly than conventional technology; managerial resistance to change; the risks perceived to be involved in adopting a new technology; questions about the profitability of new technology; the amount of capital already invested in older technology that is still competitive and will remain so for some time to come; lack of managerial and technical ability to oversee the introduction of environmental innovation; resistance of the labour force to change; and the large capital costs that are generally involved in the adoption of new technology.

Finally, rather than having a policy that actually encourages the transfer of technology, it appears that the Indian government has for some time retained policies that actively discourage technology transfer.[9] Policies of particular concern include limitations on the extent of technology transfer; delays in approving transfers; overly restrictive royalty limits; limitations on the duration of contracts; governmental controls such as mandatory sub-licensing requirements; the complexity of approval and other administrative procedures; the tax burden; customs duties; a highly sheltered import-substitution regime that has led to a lack of competitive pressure to upgrade existing technology; an industrial policy that has discouraged the development of managerial skills capable of effectively negotiating and assimilating technology; and an inadequate commercially orientated research and development policy.

New policies to facilitate the transfer of EST

To facilitate the transfer of environmentally sound technology the government of India must address these barriers. Indeed policy reforms have already taken place in the areas of technology, foreign exchange regulation, and industry and trade policy in an effort to facilitate technology transfer and foreign direct investment.[10] The new industrial policy announced in 1991 has substantially deregulated the industrial sector in India and liberalised foreign direct investment and technology imports. The reforms aim generally at shifting Indian industry from a regulatory and protected regime to a more market orientated and competitive environment. To reinforce the new industrial policy, India has taken some initiatives to reduce regulation and licensing control on foreign trade and has drafted a new technology policy that in part aims to encourage the use of environmentally sound technology.

While the reforms that have already taken place will significantly improve the transfer of technology to Indian firms, it appears that much more could be accomplished. In assessing the policies being adopted, the Indian government must clearly define its regulatory goals and must guide, stimulate and reward industry in its endeavour to achieve these goals. It is also important for the government to consider a range of measures that could be used to achieve these goals. These could include regulations,[11] economic instruments,[12] cooperative arrangements between industries, or a mix of these instruments, the choice in each case being assessed against their environmental effectiveness, economic efficiency, acceptability to industry and administrative feasibility. Each of these criteria should be examined for their effectiveness in encouraging the external acquisition of new technology and the internal absorption and diffusion of that technology. Finally, it may also be appropriate to apply different sets of policies at either the national economic level or at the level of the individual firm.

National level

Attracting foreign investment is a priority for national planners in India, and a concerted effort is under way to increase the inflow of foreign investment. Foreign investment performs three important functions in the transfer of technology. It increases capital and technology flows into the economy, creates competition, thereby improving operational efficiency, and steps up the pace of industrial growth in the economy.

Whether competition will result in the importation of the 'best available' or 'most environmentally sound' technology will depend on whether foreign multinationals bargain for terms that compromise the importation of this technology into India.[13] This in turn will depend on a number of other factors, such as the laws of the country in which the foreign firm is based, the nature of headquarters' control on environmental issues, the concern for corporate image in the host country, the environmental standards maintained by the corporation in its operations worldwide, and the presence of environmental non-governmental organisations (NGOs). NGOs have already demonstrated their ability to put pressure on both multinational corporations (MNCs) and the Indian government to improve environmental performance. For example the sharp public protest generated by NGOs against a proposed copper smelter project at Ratnagiri, which did not plan to use the most appropriate environmental technology, caught both the foreign firms involved – Sterlite Industries and Luigie Mettalurgie – and the Indian government offguard.[14]

In general, however, it seems that the involvement of MNCs will speed up the transfer of environmental and other technology. This is because the transfer of technology within MNCs themselves, from parent company to subsidiary, tends to be rapid and efficient.[15] MNCs may not, however, always be as efficient as local firms.

There will also be a need to foster more intensive domestic competition within Indian industry so as to put pressure on local firms to improve their own managerial practices rather than remain preoccupied with foreign firms as the driving force behind technology transfer. It would be better to focus on influencing the behaviour of domestic Indian firms.[16]

Joint venture arrangements are another widely used mechanism for transferring technology. However foreign suppliers tend only to supply 'current' technology to Indian firms rather than next-generation technology, which is often too costly.[17] This appears to be due to the limited relations between foreign and local firms and the advantage foreign suppliers have over purchasers in developing countries, which permits them to make contracts on terms that suit only their own interests.[18]

Environmentally sound technology is generally current or next-generation technology. New incentives are needed to encourage suppliers to transfer such technology. For example joint ventures need to be arranged between suppliers and those Indian firms that have the capacity to capture a large market share and settle into a long-term relationship. Although joint ventures in India have in the past been less attractive for foreign firms because rules have limited foreign equity to 40 per cent, it is to be hoped that current industrial policies will make India a more attractive location for these arrangements.

Of further concern has been the failure of the organisers of research and development programmes and their agencies to consider the commercial applications of their research. It is now vital that research policies consider commercial applications and that indigenous research be undertaken to develop environmentally sound technology and environmental procedures suited to Indian industry.

Individual firm level

For individual firms, factors such as enhanced technical capacity, the acquisition of technical knowledge and rewarding firms for environmental innovation have become important.

Although training arrangements, vital to any technology transfer, have been included in all technology contracts, these arrangements have been constrained by a number of factors.[19] In many cases the duration of these training arrangements are limited to the period of the transaction and cease soon after, or the training component is treated as a low priority. In others, training has concentrated only on what is provided by the main suppliers. Policies and contracts must be put in place to address these issues by lengthening, deepening and strengthening the training component of any technology agreement.

Policy instruments should also encourage firms to broaden their technical knowledge, including the skills required for operating and maintaining technology and the expertise required to generate and manage technological change. This type of knowledge can be characterised as the ability to analyse the market and its trends, to identify sources of information, and to be able to decide what technology areas to focus on. To do this firms need to ensure they employ qualified and experienced technologists; form close and extended relationships with importers to allow for the transfer of experience-related knowledge; expose their personnel to technical advances through conferences, fairs and workshops; and provide for closer interaction with research institutes, both local and foreign.

Schemes should also be devised to promote and reward innovation. Apart from stimulating technical change in a positive way, this approach would also help reduce the tendency of firms to divert resources from conventional business research and development into compliance-related research and development.[20] In addition, conventional research and development may result in innovations that go beyond regulatory requirements, while compliance-related research and development tends to use regulations as benchmarks against which to measure performance.

CONCLUSIONS

While there is no question that environmentally sound technology must be transferred to Indian industry, serious questions remain about which institutional and policy structure should be used to accelerate the transfer.

Policies must take note of and seek to redress the barriers to effective technology transfer that have been faced by Indian industry in the past. Policy making needs to be innovative and address issues at the level of both the national economy and the individual firm. The instruments used to achieve the overall policy goals may be a combination of regulations, economic incentives and cooperative arrangements within industries. The appropriate mix will depend on each policy's environmental effectiveness, administrative feasibility and economic efficiency. While regulations of the command and control type may be effective in pushing firms to shift to the best available technology, they may not provide incentives for industry to be innovative by adapting and developing new technology beyond set standards. On the other hand economic instruments have to be chosen with care, keeping in mind their relevance and acceptability to industry, their environmental effectiveness and administrative feasibility. It will also be important to ensure that operational costs are kept to a minimum by adopting approaches that use existing policy structures, self-reporting by industry and consultation.

Finally, cooperative arrangements between Indian firms and foreign firms, such as joint ventures, should be encouraged in preference to purely technical agreements, as these provide organisational modes that are more conducive to the effective transfer of technology in its broadest sense.

Notes

* The authors are with the Tata Energy Research Institute, 9 Jor Bagh, New Delhi, India. The authors acknowledge the assistance rendered by Nalini Ranganathan (TERI), Sabyasachi Mitra (*Economic Times*), Professor Deshmukh (Indian Institute of Packaging, Bombay), K. P. Niyati (CII), Dr Jurgen Wiemann (GDI, Berlin), A. Purang and Rupa Shukla of the EC Library, Dr A. Senthiwell and Dr B. Sikka (MOEF), Smita Purushottam (Brussels), M. C. Verma and Rakesh Shahani (ICRIER), Preeti Soni and M. Manikandan. In addition, the authors acknowledge further assistance from Dr R. K. Pachauri, Dr Vijay Kelkar, Veena Jha, N. Narayanan and numerous colleagues at TERI.

1. M. Bell, 'Continuing Industrialisation, Climate Change and International Technology Transfer', a report prepared in collaboration with the Resource Policy Group, Oslo, Norway, Science Policy Research Unit (SPRU), (University of Sussex, 1990).
2. C. Almeida, 'Development and Transfer of Environmentally Sound Technologies in the Industrial Sector, A Survey by the UNCTAD Secretariat' (Geneva: UNCTAD, November 1992).
3. UNCTAD *ad hoc* Working Group on the Interrelationship between Investment and Technology Transfer.
4. See S. P. Johnson and G. Corcelle, *The Environmental Policy of the European Communities, International Environmental Law and Policy Series* (London/Dordrecht/Boston: Graham and Trotman, 1989).
5. EUROSTAT: Monthly EEC External Trade: No: 07#1993.
6. There exist regulations relating to packaging in Denmark, Belgium, the Netherlands, Germany and France.
7. D. J. Teece, 'Technology Transfer and Research and Development Activities of Multinational Firms: Some Theory and Evidence', in R. G. Hawkins and A. J. Prasad (eds), *Research in International Business and Finance: A Research Annual*, vol. II (London: Jai Press, 1981).
8. Bell, 'Continuing Industrialisation', op. cit.
9. A. V. Desai (ed.) *Technology Absorption in Indian Industry* (New Delhi: Wiley Eastern Ltd, 1988).
10. Ibid., p. 27.
11. A. Warhurst, 'Environmental Regulation, Innovation and Sustainable Development', paper presented at the Third International Workshop of the Mining and Environment Research Network, UK, 14–17 September 1993, pp. 13–15, 21.
12. OECD, 'Environmental Policy: How to Apply Economic Instruments', (Paris: OECD, 1991), p. 14.
13. L. Noronha, 'Environmental Regulation, Mineral Policy and Company Approaches to Ecological Management: The Case of Goa', paper presented at the MERN Workshop at SPRU, England, 14–17 September 1993.
14. CMIE, 'Monthly Review of the Indian Economy' (New Delhi: CMIE, July 1993).
15. I. Walter, *International Economics of Pollution* (London: Macmillan, 1975), p. 138; and D. J. Teece, 'Technology Transfer and Research and Development Activities of Multinational Firms: Some Theory and Evidence', in R. G. Hawkins and A. J. Prasad (eds), *Research in International Business and Finance: A Research Annual*, vol. II (London: Jai Press, 1981), p. 46.
16. D. Scott-Kemis and M. Bell, 'Technological Dynamism and Technological Content of Collaboration: Are Indian Firms Missing Opportunities?', in A. V. Desai (ed.), *Technology Absorption in Indian Industry* (New Delhi: Wiley Eastern, 1988).
17. A. V. Desai (ed.), *Technology Absorption in Indian Industry* (New Delhi: Wiley Eastern, 1988).
18. L. Noronha, 'Policy-Making in the Indian Offshore Oil Industry with Reference to the Period 1974–1986', unpublished PhD thesis, London School of Economics, 1988, p. 201.

19. Bell, 'Continuing Industrialisation', op. cit., p. 83.
20. Warhurst, 'Environmental Regulation', op. cit., p. 22.

References

Achanta, A. N. and P. Ghosh (1993) 'Technology Transfer in the Context of Global Environmental Issues', in P. Ghosh and A. Jaintly (eds), *The Road from Rio: Environment and Development Issues in Asia*, Proceedings of the Post-UNCED Seminar on Environment Development Policy Issues in Asia, 8–10 October 1992 (New Delhi: TATA Energy Research Institute), pp. 180–92.

Bureau of Indian Standards (1993) 'Indo–EEC Trade Standards and Certification, Information for Exporters to meet the Technical the European Requirements of Market' (Delhi: Bureau of Indian Standards).

Government of India (1992) *Economic Survey 1991–92, Part I, General Review* (Delhi: Government of India, Ministry of Finance, Economic Division).

Government of India (1992) *Economic Survey 1991–92, Part II, Sectoral Developments* (Delhi: Government of India, Ministry of Finance, Economic Division).

Iba, M. (1993) 'Japanese Environmental Policies and Trade Policies: Trade Opportunities for Developing Countries', paper prepared for UNCTAD/UNDP project, INT/92/207 on Trade and Environment.

Indira Gandhi Institute of Development Research (1993) 'Trade and Environment Linkages: The Case of India', Draft report prepared for UNCTAD.

Jha, V. and A. P. Teixeira (1993) 'Are Environmentally Sound Technologies the Emperor's New Clothes?', paper prepared for UNCTAD/Government of Norway workshop on Transfer and Development of Environmentally Sound Technologies, Oslo, 13–15 October.

Johnson, S. P. and G. Corcelle (1989) *The Environmental Policy of the European Communities*, International Environmental Law and Policy Series (London/Dordrecht/Boston: Graham and Trotman).

Menon, K. S. V. (1991) *Technology Transfer – Concept. Modalities and Case Studies* (Delhi: Goldline Publishers).

Sridharan, E. (1991) *Liberalisation, Self-Reliance and Technology Policy in India in the Eighties: A Shift in the Terms of Discourse* (Delhi: Centre for Policy Research).

18 Trade, Environment and Development Cooperation

Ebba Dohlman*

The Development Assistance Committee of the Organisation for Economic Cooperation and Development gives priority to the policies of trade, environment and development cooperation, and in particular focuses on the role that aid agencies can play in promoting compatibility between these policies. In addition the committee is seeking ways to assist developing countries to use new trading opportunities at the interface of trade and the environment and overcome trade constraints that may result from increasingly stringent and complex environmental regulations. Finally the committee, and particularly its working party on development assistance and the environment, is taking steps to ensure that the findings of the many different fora working on trade and the environment are incorporated into the development cooperation strategies of donor nations and operationalised in the planning, design and implementation of aid programmes.

POLICY COHERENCE BETWEEN TRADE, ENVIRONMENT AND DEVELOPMENT COOPERATION

Promoting policy coherence and compatibility between trade, the environment and development cooperation was the principal objective agreed upon at the United Nations Conference on Environment and Development and laid down in the Rio Declaration and Agenda 21.

However the actual role of trade and trade policy in promoting or undermining sustainable development is still not fully understood. Most experts agree that, in general, the direct effects of trade on the environment are limited, as only a small share of environmentally sensitive goods enter into trade and because trade is only one of many factors affecting the environment. Much depends on the specific context, such as a high level of dependence on one or a few commodities for export earnings, as well as the different capabilities nations have to implement appropriate economic policies and effective environmental protection regimes.

A number of developing countries are engaged in a successful process of trade liberalisation and a general deregulation of their economies. This is a positive trend because trade has usually been a major factor underlying a nation's development. From an environmental perspective, open trade also encourages more efficient use of resources and promotes the transfer of technologies that can improve environmental management. The increased competition from trade improves innovation and productivity in resource use. Although trade liberalisation cannot by itself guarantee good environmental management, it promotes the conditions necessary for ensuring that good environmental policy can be effective.

One of the keys to sustainable management of natural resources is implementation of the proper economic incentives in producer countries. Economic instruments, standards and regulations can all be used to ensure that environmental costs are adequately reflected in the production process or the consumption of natural resources.

Trade policies can also provide environmental incentives, but only if they improve rather than restrict access to import markets. In addition these policies must ensure that the maximum value can be added to sustainably produced goods. In this context, the high level of tariff barriers for processed goods in developed countries relative to unprocessed goods is of particular concern to developing countries. Although it is not yet clear how the environmental impact of more processed goods compares with that of basic commodities, tariff escalation in developed countries will reduce market access for processed goods. Moreover tariff escalation will inhibit economic diversification in developing countries, retard entry into the export-oriented processing industries that are most accessible to developing countries, pressure producers of natural-resource-based commodities to intensify their exports, and lead to overdependence on unprocessed commodities. Experience also shows that an increased supply of commodities frequently leads to lower world prices. This may set off a downward spiral of decreasing revenues for the countries involved and an inability to invest in sustainable production methods.

In some cases an incomplete understanding of the implications of harnessing comparative advantage to make gains in trade can lead to policy incoherence. Some environmental experts and economists hold the view that lower or non-existent environmental standards in developing countries can be exploited as a comparative advantage. This has led to the export by some developed countries of hazardous waste in order to avoid the costs of properly disposing of them at home, and in other cases the export of goods that are no longer allowed or desirable in domestic markets, including harmful pesticides and insecticides. Some progress has, however, been made recently in this respect with agreement that the Basel Convention would be extended to cover a complete export ban of hazardous waste from OECD countries to developing countries by 1997.

Developed countries themselves have expressed concern about environmental dumping by developing countries. 'Environmental dumping' is defined as the selling of goods in developed-country markets at prices that do not reflect the environmental damage caused during their production in the exporting country. However the concept of environmental dumping also represents a misunderstanding of comparative advantage. There are a host of factors besides the environment – such as the cost of labour – that result in lower prices for exports from developing countries. Proposals by developed countries to apply countervailing duties because of the 'implicit subsidies' in the price of goods that do not reflect the environmental damage caused during their production represents a set of incoherent policies. Putting these policies in place would result in a potentially spiralling set of trade restrictions and tensions that would not contribute to environmental protection or sustainable development.

Aid agencies can play an important role by ensuring policy coherence so that the negative economic effects of developed country environmental and trade policies do not negate the positive economic benefits of development assistance. Although aid agencies may have limited ability, at least in the short term, for overriding the powerful interests that lead to more distortive trade policies, they nevertheless have an obligation to encourage a more integrated, coherent policy approach through interministerial or interagency coordination and dialogue with the private sector.

EMERGING OPPORTUNITIES FROM 'GREEN' TRADE

The demand for environmentally friendly goods in OECD country markets and in certain developing country markets has increased enormously over the past decade. The International Finance Corporation expects that the market for environmental goods will grow from its current $300 billion to $600 billion by the year 2000, partly as a consequence of the expanding range of environmental regulations. While this demand may present certain adjustment problems for developing-country exporters, they may also provide new opportunities. Aid agencies could help developing countries to benefit from the 'greening' of world markets.

Developing countries are a potentially rich source of environmentally friendly products and technologies. Tropical forests are a well-known source of medicinal plants. Export products from developing countries are being produced increasingly in organic or environmentally friendly conditions. For example products such as rice, cotton and other fibres, bananas, fruit, honey, flowers, biocides, oils and fats, vegetable tanning of leather, coffee, tea and cocoa are increasingly being produced in an environmentally friendly way.

However, even when these products can be identified as environmentally friendly, developing countries are frequently not in a position to exploit them on the international or even the domestic market. Some non-governmental organisations, such as TWIN and TWIN TRADING, the Max Havelaar Foundation, the Fairtrade Foundation, the Rainforest Alliance and the Organic Chocolate Manufacturers, have tried to fill the gap between product identification and promotion in international markets. Some private firms, such as the Body Shop, Ben and Jerry's Ice-cream, and l'Occitane in France, are also promoting environmentally friendly, developing-country products through 'trade not aid'. These groups are providing useful avenues for linking the environment with trade and development. However their work alone is not sufficient to meet the need or the demand. Aid agencies could usefully build on this work and assist in the promotion of environmentally friendly goods, particularly by helping developing country producers to obtain eco-labels, gain mutual recognition for their own standards and promote consultations. The OECD Development Assistance Committee itself could also bring together profit and non-profit organisations to share experiences concerning environmentally orientated trade promotion, among other things.

Some donors are already examining ways they might improve the image of developing country products that have an environmental advantage. France and Germany have for instance introduced, or are in the process of introducing, eco-certification for timber in consultation with developing-country producers. Fair trade marks are being awarded by some NGOs to coffee and other products that are produced in an environmentally friendly way. The Rainforest Alliance has developed a code of environmental standards for banana producers and awards an 'ECO-O.K.' seal of approval.

Governments and ecological interest groups tend to support eco-labelling schemes since the promotional effect of a label sets incentives for producers to improve the environmental qualities of their products, and may help meet environmental objectives. Moreover labelling is a market-oriented instrument that does not establish any binding requirements or bans. However eco-labels call for an overall assessment of the ecological impact of a product during its life-cycle, even though there is no general agreement yet on how to weigh different types of environmental impact, or how to evaluate the net or total environmental impact of a product. Indeed there is a risk that if too many different schemes emerge in the same sectors, each with its own definitions and criteria, this will undermine the usefulness and credibility of the schemes, cause confusion, possibly discrimination and lead to income foregone in many developing countries.

A ROLE FOR AID AGENCIES

Aid agencies can play a role in helping developing countries to promote their environmentally friendly exports and seize opportunities arising from the 'greening' of global markets. They also have a role in helping developing countries to overcome some of the potential obstacles associated with environmental regulations in developed countries, for example by providing access to timely and comprehensive information. The OECD will also have a role in making recommendations about how developing countries can minimise the adverse trade effects of these environmental measures.

Financial and technical assistance could also be offered by aid agencies to developing countries in a number of areas, including the following:

1. Linking trade promotion activities to the environment.
2. Providing training for environmental impact assessment or for the assessment of a product's environmental performance.
3. Encouraging cooperation among producers to attain economies of scale in production and for access to technology.
4. Establishing links between producers and consumers, both within developing countries and internationally.
5. Disseminating information to producers about changing trends in markets and new environmental quality standards and to consumers about the benefits and qualities of environmentally friendly products.
6. Developing and promoting eco-labelling schemes and encouraging consultations among producers, exporters, importers and consumers about their standards.
7. Encouraging mutual recognition of labels, promoting training for verification, certification and other quality controls and setting up certification centres or testing facilities where these are not already adequate.

CONCLUSIONS

Aid agencies have an important role to play in making trade, environmental and development cooperation policies mutually compatible. They also have a role in assisting developing countries take advantage of the new trading opportunities arising from concern in developed-country markets about the environment or avoid some of the trade restrictions being imposed for environmental reasons. Of particular importance will be the provision of timely and comprehensive information about environmental opportunities and regulations, and financial and technical assistance. While the catchphrase of developing countries in the past few years has been 'trade not aid', there is still an important role for aid agencies and aid programmes in helping developing countries to meet their potential as trading nations.

Note

* This paper reflects the views of the author and not necessarily those of the OECD.

19 Principles for Making Trade and the Environment Mutually Compatible

Nevin Shaw

The environment and the economy are so inextricably interrelated in their contribution to human well-being that environmental management must become an integral part of economic policy making, including trade.

Indeed we urgently need to locate trade and environment in the broader context of finding new ways of integrating domestic and international economies and environments in our decision making. This should be done with a clear commitment to recognising our common but differentiated responsibilities for the past and future use of planetary resources. In addition, issues of equity, poverty alleviation and lack of capacity have to be addressed. To assist the global community to achieve these objectives, the International Institute for Sustainable Development (IISD) has developed a set of trade and sustainable development principles.[1] Only when this wider set of issues is addressed effectively will our efforts lead to sustainable development.

ECONOMIC REFORMS FOR SUSTAINABLE DEVELOPMENT

In domestic terms, budgets must be prepared with the overall objective of achieving long-term sustainable development. This requires an effective macro- and microeconomic policy framework.[2] Microeconomic policy reforms need to be based on considerations of the efficiencies to be gained by good environmental management, such as efficient resource use through waste minimisation and energy conservation. Consumers often contribute to this process by demanding more environmentally friendly goods and services, but investment in these types of products and processes must also be made more attractive.

Specifically, microeconomic policy reforms should gradually, but fully, place the costs of depletion and pollution on the producers and consumers who currently benefit from these practices. Where price controls are in place, additional measures are likely to be necessary, such as raising the price of a traded good to the international level. It is also necessary to define and assign enforceable property rights over natural resources to encourage their optimal use. In addition there should be a gradual shifting of environmental management from dependence on command and control regulations, such as emission standards and mandated production methods, to economic instruments such as taxes, deposit-refund schemes and grants. Furthermore there could be better use of incentives and disincentives to spread rather than concentrate economic activity spatially. Environmental impact assessments and similar mechanisms can be used to educate the public and persuade people to agree to pay for the consumption of environmental resources.

It is clear, however, that there is a need to avoid the premature use of punitive trade policies to enforce measures to deal with the irreversible damage being done to the planet.

Nevertheless, as noted above, there must be a major restructuring of the economy if we are to avoid continual environmental degradation. Implementation of these microeconomic reforms will need to be prudent and predictable, as well as being phased in so as to maintain the confidence of investors. The adjustment to sustainable forms of development will require equitable burden sharing through social security allowances and skills upgrading. Although this kind of restructuring will no doubt cause some economic and social disruption in the short term, these measures cannot be put off until the future.

Measuring and monitoring what activities are sustainable or unsustainable will also require investment in the development of new kinds of statistics, analysis, more open policy-making processes and monitoring and enforcement capabilities. These cannot, however, be taken for granted. Indeed the experience with environmental impact studies indicates how ambiguous they can be and how much caution is needed when choosing between alternative policy choices.

CANADIAN EXPERIENCES WITH NAFTA

Canada is perhaps one of the few countries that has had to integrate trade and environmental policies in a practical context. The changing views of the public towards sustainable development, as well as the new perceptions of the relationship between the economy and the environment, came together with trade as Canada, Mexico and the United States were in the middle of negotiating the North American Free Trade Agreement (NAFTA). The policy communities dealing with trade, the environment and economic development, who have in the past operated in relative isolation from each other, now had to cooperate to arrive at a consensus during these negotiations. Their own cultures and operating styles made policy integration all the more difficult. Moreover the complexity of the various issues raised and the lack of adequate information only increased the scale of challenge in moving NAFTA forward coherently.

Several steps were taken to facilitate the progress. For the first time representatives of environmental interest groups were appointed to a Non-Governmental International Trade Advisory Committee and to eight Sectoral Advisory Groups on International Trade, thereby opening up the policy process. In addition a decision was made directly to address trade-related environmental concerns. Environmental considerations were made an integral part of the negotiating mandate of each of Canada's lead negotiators. The Environment Department was also represented on Canada's team of negotiators.

These steps were taken to ensure that environmental considerations were heard during all stages of the negotiations. The result is a major trade agreement that directly responds to many environmental concerns, such as standards, dispute settlement, and the relationship between trade agreements and multilateral environmental agreements.

An environmental impact assessment (EIA) of NAFTA was also undertaken to document its potential effects. The EIA was conducted by a newly established Interdepartmental NAFTA Environmental Review Committee, which was regularly briefed and consulted by the negotiators and which itself consulted with provincial governments, academics, business, environmental and labour leaders.

The agreement established a Commission for Environmental Cooperation, with a potentially large enough mandate to address shortcomings in the agreement and promote new possibilities for sustainable development. It is hoped that the Commission will be supported by the multi stakeholder national and provincial round tables that are being promoted by IISD to partner and promote dialogues in sustainable development.

In sum, trade and the environment can be made more mutually compatible through this type of opening up of the policy-making processes to enable decision makers to adapt and correct mistakes on the basis of dialogue, deeper understanding, acceptance of each others' objectives and cooperation. While NAFTA is instructive in this respect, other multilateral institutions first need a set of interlinked principles by which their basis of understanding can be improved, common ground found and possibly new rules and agreements developed to enhance the relationship between the environment and the economy.

IISD PRINCIPLES

As noted earlier, the IISD released a set of these principles in 1994, after having incorporated feedback from hundreds of contacts world-wide from the trade, environment and development policy communities. The IISD also brought together nine experts from widely disparate backgrounds and countries and asked them to seek agreement on a set of principles that all nations could apply to ensure that trade, environmental protection and development are pursued more harmoniously.

The principles may be described as efficiency, equity, environmental integrity, subsidiarity, international cooperation, science and precaution, and openness. They provide a mechanism that is flexible and responsible in addressing these issues as well as acknowledging the complexity and uncertainty of trade and environment interlinkages and the unacceptable tendencies of some to choose certain actions while ignoring others. The principles seek to enhance the ability of different countries to negotiate and adhere to trade/environment agreements as they promote wider concerns among governments. What follows is a brief description of each principle and how it applies to the trade/environment debate.

(1) *Efficiency/cost internalisation.* Efficiency and cost internalisation entail the progressive and predictable reflection of unpaid environmental costs in the price of products so as to minimise inputs per unit of output and promote progressive trade liberalisation. Environmental costs can be internalised through the combined use of economic instruments and various forms of environmental regulation. They can also be reduced through the elimination of environmentally damaging subsidies and escalating tariffs. The World Trade Organisation is now examining the use of taxes and charges in this context.

(2) *Equity.* Further trade liberalisation and the sharing of capital, knowledge and technology is needed to promote greater equity today and tomorrow (within and between countries, as well as groups of countries) given past, present and expected use of the environment. As standards of living outside the OECD are far too low (poverty is a key result of environmental degradation), future policy must build in greater equity considerations. This includes assistance to more coherently organised institutional structures that permit sustainable development issues to be articulated and

debated, and to build an institutional capacity for formulating and implementing sustainable development policies. It also includes additional market access to allow poorer countries to include environmental costs in their export prices. Barriers to imports of poor countries should be eased.

(3) *Environmental integrity.* This refers to measures that respect and help maintain environmental integrity where efficiency and cost internalisation cannot capture environmental values, for example where irreversible depletion of fisheries or forests are involved, or irrevocable losses such as extinction of species. Where such efforts require the use of GATT-inconsistent trade measures, they should be within the context of prior multilaterally agreed criteria (for example CITES and the Montreal Protocol).

(4) *Subsidiarity.* Subsidiarity is the implementation of environmental measures at a domestic jurisdictional level appropriate to the source and scope of the problem and appropriate to effectiveness in achieving objectives. Where there are significant transborder impacts, there should be international cooperative efforts.

(5) *International cooperation.* More aggressive efforts should be made to develop common understanding and approaches as to how our multilateral system should evolve and how questions relating to a variety of issues could be addressed, including more open, equitable, effective and alternative dispute settlement processes.

(6) *Science and precaution.* We should encourage a precautionary approach to the adoption of environmental policies that would allow governments to deal with serious threats of environmental harm in advance of conclusive scientific evidence concerning that harm. Such polices should be adapted as new scientific evidence becomes available. Openness in this approach is essential (for example climate convention and ozone-depletion issues).

(7) *Openness.* Timely, easy and full access to information by all affected or interested parties, and public participation and accountability in the decision-making process informed by considerations of feasibility.

At present many deficit and debt-ridden governments do not share the environmental priorities of the United States and the European Union, nor views on how these priorities are to be given effect. This is reflected in their position on Geneva's trade rule-making processes and the subsequent results. The principles give legitimacy not only to improved environmental protection, but also to the related issues of equity and cooperation in ways that cannot be ignored by major governmental powers. Indeed it is hoped that the principles will influence the creation of new coalitions for change, publicly linking expected change in behaviour to incentives as well as creating pressure on laggards, including pressures to avoid the use of trade measures, while at the same time addressing the demands of global equity.

The principles also provide a basis for negotiations and accountability in a variety of activities that support trade and sustainable development. They can improve the quality of multilateral environmental agreements. They require information sharing and regular interactions between interested and affected parties on a common set of issues, thus making commitments more credible. Indeed in this regard they would support notification measures, trade policy reviews and the dispute settlement processes of the World Trade Organisation. They accommodate minimum international standards and agreements, eco-labelling, capacity building and even the use of trade measures within the larger context of achieving progress towards sustainable development.

There is also a need to build capacity to negotiate environmental commitments that respond to stakeholder concerns domestically and internationally. Information, research, analytical, political, legal, administrative and technical competencies will be the keys to integrating policy in this way. In this regard, non-governmental and governmental organisations must find more creative means of helping the vast majority of poorer nations. Increased or redirected bilateral and multilateral assistance, increased market access on a preferential basis, debt relief, concessionary export credits, greater flows of private foreign investment and better commodity agreements are examples of what is required. The lending programmes of the World Bank and other international financial institutions should also be made more accessible. Other institutions should look to funding the transfer of technology, research and development, as well as the joint implementation of projects.

The IISD principles also require packaging and labelling programmes to be more open to foreign participation, and support aggressive promotion of the exports of poorer countries by the OECD and other groups of countries. This approach respects the growing judgement that a planetary increase in production and consumption to meet the expected several-fold increase in demand is simply unsustainable. It also more collectively and equitably addresses the need to alter production and consumption patterns and reallocate resources.

The IISD expects the various institutions involved in trade and environment interlinkages to adapt and work together in a pluralistic approach to future problem solving. The IISD intends to apply the principles to the operations of the GATT/WTO, NAFTA, the European Union, the World Bank and the International Monetary Fund, and to the operations of multilateral environmental agreements.

Twenty-first century behaviour must be based on agreed principles for the benefit of all rather than the exercise of power.

Notes

1. IISD, *Trade and Sustainable Development Principles* (Winnipeg: IISD, 1994).
2. *Making Budgets Green* (Winnipeg: IISD, 1994).

20 Conclusions and Policy Recommendations

Veena Jha

When framing policies for making trade and the environment mutually compatible, it is important to understand the complexity of their interlinkages, particularly when they are juxtaposed against development. Trade liberalisation may under some circumstances be beneficial to the environment by, for instance, improving resource allocation, but under other circumstances it may exacerbate existing environmental problems. Similarly, environmental rules in developed countries may constitute barriers to trade and imply onerous adjustments for developing countries in some instances, but in others they may provide an opportunity to improve the environment and simultaneously gain trade benefits. This book has demonstrated that the question of whether trade liberalisation is beneficial to the environment or whether environmental policies generate significant trade effects is ultimately one that is empirical in nature. The important question is not whether these impacts will or will not arise, because they certainly will, but how quantitatively important they are in practice.

The studies in this book have also shown that there are significant differences between the perceptions and institutional capacities of developing as compared with developed countries when considering the implementation of environmental policy. In particular, developing countries lack a general administrative and monitoring capacity and have only a limited ability to enforce policy instruments. The studies have demonstrated that non-protectionist options should be favoured and that production externalities are best tackled with incentive-based mechanisms that minimise enforcement costs. Moreover it is to be hoped that a better understanding on the part of those in developed countries about the process of green policy formulation in developing countries will help preempt green protectionism and foster environmental cooperation.

Infrastructural investment may be one of the most important pro-environmental process-related improvements that developing countries can make. It seems clear that in developing countries not only does industrialisation and its accompanying urbanisation quickly increase pollution of the environment, but also that these changes usually overwhelm the capacity of the national infrastructure and the ability of authorities to mobilise resources for prevention and control programmes. Consequently, policies that aim to promote improvements in infrastructural facilities will also simultaneously mitigate pressing environmental problems.

Regarding environmental factors, market access and competitiveness, the studies indicate that in most sectors there are no major environmental requirements for traded products. However, where requirements do exist, the impact on trade and the competitiveness of developing countries can be serious. Moreover developing countries fear that environmental factors may become increasingly important in the future and develop into significant barriers to trade. Among the difficulties identified by developing countries in meeting the environmental requirements of OECD countries are the high costs of adaption; the irrelevance of foreign standards to local conditions; the lack of timely and adequate information and consequent transaction costs; the difficulties in understanding the requirements as well as testing for and monitoring them; the perceived lack of scientific data for specific threshold or limiting values and the uncertainty that arises from rapidly changing requirements in overseas markets. More importantly, the uncertainties attached to obtaining premiums for making environmental improvements have deterred a number of developing countries from investing in the expensive equipment and processes required to meet the environmental requirements of OECD countries.

Some of the fear experienced by developing countries could be alleviated by improving their market access conditions. Improved market access, which is essential for the economic development of developing countries and countries in transition, has an important role to play in moving the global community towards sustainable development. Improved market access can provide the resources for environmental improvement and increased efficiency as well as facilitating diversification of production and exports, thereby reducing the heavy dependence of many developing countries on a few commodities for their foreign exchange earnings. While the implementation of the results of the Uruguay Round will substantially improve market access, it remains important to identify areas where further trade liberalisation, including the reduction or elimination of tariff escalation, can contribute to sustainable development.

Another deep concern for developing countries is whether premiums in price will accrue from the investments they make in environmental improvement. Empirical evidence to date does not conclusively prove that these premiums are significant. Price premiums paid for the environmental quality of a product depend on many factors, such as the availability of close substitutes, consumer tastes, fashions and the association in the consumer's mind between the product and its environmental attributes. Further research is required to identify when premiums are more likely to accrue and when they are not.

Nevertheless, even if market premiums are not obtainable, in certain cases investment in the prevention of pollution may yield positive rates of return similar to non-environmental investments. However the case-studies in this book demonstrate that investment in pollution prevention and abatement may be especially onerous for small-scale producers. Cleaner as opposed to end-of-pipe technologies, for instance, may require a minimum scale of operation in order to be economically viable, even though operating costs may decrease over the long term.

While differences in the costs of compliance with environmental regulations between small and large firms may be obvious in the case of capital costs, they also tend to be reflected in running costs. Unlike large firms, for which internally generated sources of finance may be available, small and medium-sized enterprises may find it difficult to gain access to capital for investments due to low or negative private returns. If the economies of scale are significant for the particular environmental investment, this will affect smaller firms more than larger ones. It might also be argued that small and spatially dispersed sources of pollutants need not be approached with the same sense of urgency and the same type of regulatory instruments as pollution originating from large and spatially concentrated sources. However, where small firms are concentrated in the same area, government assistance or collective initiatives may be required to improve environmental conditions. Aid agencies, for example, might investigate mechanisms that facilitate environmental improvement for small-scale firms, including the financing of common effluent treatment plants, better information dissemination and research into cost-effective solutions for small-scale operators. In light of these differences, there may be a need to grant time-limited exemptions for smaller firms in the course of enforcing stricter environmental standards for an industry as a whole.

Technical assistance and cooperation in capacity building should be used to help developing countries to design cost-effective regulatory instruments and upgrade and enforce their process standards. It should be noted that the private sector has an important role to play in this context. Research has demonstrated that institutional weaknesses in developing countries may preclude effective monitoring of environmental standards. Consequently, further investigation into the role of mechanisms that internalise externalities, but which do not put too much stress on the monitoring capacities of governments, is required. Alternatives such as self-certification, market checks, competitors' peer assessments and techniques for verifying environmental quality without recourse to extensive testing and administration also need to be appraised.

Government-to-government technical assistance will also be of significant benefit to developing countries, at least initially. The examples in this book of governmental assistance provided to India for the adjustments made to new environmental standards in the leather industry demonstrate just how important this traditional source of assistance remains.

The importance of harmonisation, especially at the regional level, has also been a common theme in the chapters of this book. The possibilities for harmonisation deserve more detailed scrutiny, with future work perhaps focusing on the problems of harmonising environmental approaches between economically disparate but geographically proximate countries.

Finally, it is clear that the transparency of environmental measures must be improved. Transparency not only refers to the timely dissemination of information, but also to the effective participation of developing countries, both in formulating environmental measures at the national level and in discussing trade and the environment at international meetings. Participation can be encouraged in several ways. For instance there is room for more training about how environmental standards are set in developed countries. Managers and technicians from developing countries would also benefit from learning about and experimenting with various instruments to find out which work and which do not in the context of developing countries. Building awareness and providing education on the different aspects of trade, environment and development is also an important aspect of capacity building. Aid agencies might also explore the possibility of promoting mutual recognition of environmental requirements and standards, acknowledging of course, that achieving similar environmental conditions may require distinct initiatives in different parts of the world, depending on a particular stage of development.

The various paths being explored by the international community to resolve the difficulties brought about by the interlinkages between trade and environmental policies is of vital concern to developing countries. It is important that the experience of developing countries with trade and the environment be understood fully, not only by those from developing countries themselves, but also by those from developed countries. It is hoped that this book will go some way toward shedding light on that experience.

Index

Abbreviations used in subheadings: EST...environmentally sound technology; PPM...process and production methods.

Agenda 21, UN Conference on Environment and Development 16, 31, 37, 49, 187, 203
Agricultural and Processed Food Products and Export Development Authority (APEDA) 135
agriculture and food industry
  India 125, 134–6
  Japanese environmental laws 193
  seafood 125, 135–6
aid agencies 205, 206, 207–8, 219
airconditioning 127–30, 191–2
Almeida, C. 186
Asia Pacific People's Environment Network 88
Asian Forum of Environment Journalists 88
Asian Wetland Bureau 88
Association of Textile Producers, MST and MUT labels 56–9, 132, 189
automobile industry, India 189–90

Bangladesh 83, 84–5, 86, 87, 97–103
  environmental degradation 83, 84–5, 86, 87, 101–3
  leather industry 97–103
  pollution control 101–3
Bangladesh Pesticide Ordinance 87
Basel Convention 205
Bata Shoe Company (Bangladesh) Ltd 99
Bell, M. 186
Ben and Jerry's Icecream 206
best available process technology (BPT) 133
best available technology (BAT) 133
Bhutan 83, 105–7
Biocon, India 190–1
Biodiversity Convention 87
Body Shop 53, 206
Bureau of Indian Standards (BIS) 133
Business Council for Sustainable Development 187
'buyer pays principle' 16

Canada, integration of trade and environmental policies 210–15
carpet industry, Nepal 145, 146
Celanese Mexicana 175
Central Environment Agency, Sri Lanka 165
Central Pollution Board, India 133
certification, environmental 2, 35–6, 38, 191, 192
chemical industry, India 190
child labour 131
Child Labour Prohibition Act (1986), India 131
China, and Montreal Protocol 173
chlorofluorocarbons (CFCs) 114, 123, 127–9, 174, 191
coastal environmental damage 85
Commission for Environmental Cooperation, Canada 211
Commission on Sustainable Development (CSD) 19
Committee on Economic Cooperation (CEC), SAARC 90–1
Committee on Studies for Cooperation in Development in South Asia (CSCD) 89
comparative advantage
  and environment 16–17, 74–6, 109–10, 117, 162, 204–5
  see also competitiveness of developing countries
competitiveness of developing countries 41–9, 69–76, 106–7, 119–20, 125–7, 177

competitiveness – *continued*
and eco-labelling 17, 46–7, 119, 177
and environmental policies/standards 5, 27–9, 41–9, 69–76, 119–20
and product-specific policies 41–4
and relocation of polluting industries 71–4
see also comparative advantage
compliance costs
country differentials and subsidies 28–9, 30–1
of environmental standards 25–6, 28, 45, 48, 69, 127, 219
consumer interest groups 52–3
see also 'green consumerism'
costs
environmental standard compliance 25–6, 28, 45, 48, 69, 127, 219
recycling 44–5
transaction 45, 48, 127
cotton industry 150–1, 152, 189
'eco-cotton' 150–1, 189
see also textile industry

deforestation, South Asia 82–3
developing countries
appropriateness of OECD environmental criteria 17–18, 34, 47, 55–6, 92
competitiveness of 41–9, 69–76, 106–7, 119–20, 125–7, 177
eco-labelling by 55–6, 65–6, 127
effect of eco-labelling on 17–18, 47, 119, 124, 136–8
environmental protection in 16–17, 74–6, 86–8, 92, 107
impact of packaging regulation 119, 145, 159, 192
learning from examples of OECD countries 160
and markets for 'green' products 5, 51–2, 105, 138, 143, 145–6, 205–7
problems in meeting environmental standards 123–4
trade liberalisation 2–3, 16, 90–2, 162, 163–4
Development Assistance Committee, OECD 203, 206
development cooperation, trade and environment 203–8
development policies
and environmental policies in South Asia 81–92
for sustainable development 209–10
DKK Scharfenstein 129
dyestuffs 17, 132–4

'eco-cotton'
India 189
Pakistan 150–1
eco-dumping 30–1, 205
eco-labelling 33–5, 46–7, 53–9, 63–4, 65–6, 136–8, 214
by developing countries 55–6, 65–6, 127; India 16, 131, 135, 137
of dyestuffs 17, 132–4
effect on trade of developing countries 2–3, 17–18, 47, 119, 124, 136–8, 159
German Blue Angel system 5, 47, 51, 53–7, 63, 127
as 'green imperialism' 18
harmonisation 58, 119, 124
IISD principles 214
Indian Eco-Mark 131, 135, 137
Japanese Eco-Mark 194
in leather industry 119
and market access 27–8, 46–7, 58–9, 177
mutual recognition of 34–5, 38
and PPM standards 33–5, 47
and product life-cycle 206–7
of textiles 3, 47, 63–4, 137–8, 188–9; German schemes 51, 56–9, 132, 189
Eco-Mark
India 131, 135, 137
Japan 194
economic instruments, use in environmental policies 43–4, 52, 210

'end of pipe' technology 113, 119, 174–5, 179
energy conservation 42, 53
entry barriers
  eco-labels as 58–9
  environmental standards as 18–19, 21, 81, 92, 117, 119
  Technical Barriers to Trade (TBT) Agreement 24, 28
  see also market access; protectionism; trade restrictions
environment and trade 19–20, 163–4, 203–8
  Canada 210–15
  in South Asia 2–3, 81–92
  see also trade–environment mutual compatibility
environmental auditing 16
environmental degradation
  and development programmes 81
  and industrialisation 156, 165–6, 217–18
  and poverty 17
  in South Asia 82–6
environmental impact assessment 210
  in India 16
  of NAFTA 211
  in South Asia 87
  in Sri Lanka 87, 158, 165, 166
Environmental Impact Assessment Act, Sri Lanka 166
environmental policies
  effect on competitiveness 41–9, 69–76, 106–7, 119–20, 125–7, 177
  in Germany 51–66
  global or local 172–4
  and packaging 44–6, 48, 51–66, 127
  product 41–9
  product-specific 41–4
  and South Asian development policies 81–92
  use of economic instruments 43–4, 52, 210
environmental protection
  and comparative advantage 16–17, 74–6, 162, 204–5
  consumer attitudes to 52–3
  in developing countries 16–17, 74–6, 86–8, 92, 106–7
  and 'green' protectionism 17–19, 23, 64, 66, 117
  international agreements on 16, 30, 120, 126, 172–3
  in South Asia 86–8, 92, 93–4 (*Appendices*)
environmental standards
  and competitiveness of developing countries 5, 16–17, 69–76, 106–7, 119–20, 125–7
  compliance costs 25–6, 28, 45, 48, 69, 127, 219
  country differentials 28–9, 30–1, 153–4
  in developing countries 123–4
  harmonisation of 18–19, 29–31, 188, 220
  ineffectual monitoring 87, 114, 217, 219
  minimum international 30
  in OECD countries 125–7, 176–7, 218
  quality assurance as 23
  and relocation of production 69–76
  testing facilities 122, 131, 136, 138
  as trade barriers 18–19, 21, 81, 92, 117, 119
  and trade liberalisation 3, 123
  see also process and production method (PPM) standards; product standards
environmentally friendly products
  certification of 2, 35–6, 38, 191, 192
  from developing countries 5, 51–2, 105, 138, 143, 145–6, 205–7
  market for 123–4, 205–7
environmentally sound technology (EST) 110–11, 113–114, 171–81, 185–201
  best available 133, 197
  choosing 113–14, 129
  cost problems 110–11, 114, 178–9, 195
  defined 187

environmentally sound technology – *continued*
'end of pipe' or clean production 113, 119, 174–5, 179, 219
essential for sustainable development 180–1, 185
Karnal technology 114
market in OECD countries 175–6
non-CFC 129–30, 174
see also technology transfer
equivalencies, eco-labelling 33–5, 38
ESCAP 153
European Union (EU) 33, 57, 163, 188–92, 214
eco-labelling 33, 57
Environmental Policy (1972) 188
environmental regulation 188–92
Export Promotion Bureau, Pakistan 150

fair trading 53, 65, 206
Fairtrade Foundation 206
Food Sanitation Law, Japan 193
footwear industry 99–100, 136–7
foreign direct investment, in EST 179–80, 196–7
forests and forestry industry *see* timber industry and forests
Framework Convention on Climate Change 87
France 61, 206

German Federal Environmental Agency (FEA) 54, 66
German Institute for Quality Control (RAL) 54–5
German Society for Promotion of Partnership with the Third World (GEPA) 65
German Society for Technical Cooperation (GTZ) 56, 66
Germany 5, 51–66
ban on use of PCPs 42, 120–2, 125, 177, 189
Blue Angel eco-labelling system 5, 47, 51, 53–7, 63, 127
eco-certification of timber 206
environmental policy making 52–3
Green Dot and Dual System (DSD) 5, 51, 60–1, 151
Packaging Ordinance (1986) 5, 45, 51, 59–63, 127, 131, 192
recycling 45, 46, 60–1
Gesamttextil e. V. 56–9
Global Clean Water Incentives Act, US 173–4
global commons 172–3
Global Environmental Facility (GEF), Montreal Protocol 32
global warming 86, 119
Gokak Mills 189, 192
Good Manufacturing Practices (GMP) 191
'green consumerism' 27–8, 33, 59, 143, 145–6, 150
'green imperialism' 18, 122
'green protectionism' 4, 64, 66, 217

harmonisation
of eco-labelling 58, 119, 124
of environmental standards 188, 220
packaging regulation 45
of pollution regulation 18–19
of PPMs 29–31
South Asian 90–1, 92
hazardous substances 42, 45, 125, 157–8, 190, 204–5
Hindustan Insecticides Ltd 135
Hong Kong 176

incentives
for EST transfer 196, 197–8, 199
for sustainable development policies 204, 210, 217
India
Central Pollution Board 133
child labour 131
effect of packaging regulation 192
environmental degradation 83, 84, 85
environmental policies 16, 87, 125; eco-labelling (eco-mark) 16, 131, 135, 137; Montreal Protocol 110, 123, 128; sustainable development 124–5
hazardous substances: pesticides 114, 125, 130–1, 134–5, 193;

use of CFCs 114, 127–30, 191–2; use of PCPs 42, 120–2, 125, 177, 189
industries of environmental interest: agricultural and marine products 134–6, 193; automobiles 189–90; chemicals 190; leather industry 42, 117–22, 125, 136–7, 189; pharmaceuticals 190–1; tea production 130–2; textile industry 137–8, 188–9; timber industry and forests 17, 112–13, 131–2
joint ventures 197–8
refrigeration 128–9, 191–2
trade liberalisation 16, 111–13, 120
trade–environment compatibility 109–14, 117–20, 123–39
transfer of EST 187–99; at firm level 198–9; at national level 196–8; barriers to 195–6; effect of EU environmental regulation 188–92; effect of Japanese environmental regulation 193–4; and ISO standards 194; policy framework for 194–9
Indian Plantation Labour Act (1951) 131
Indian Tea Research Association 130
Indo–German Export Promotion Project 122
Indonesia 61
industrial migration *see* relocation of production
Industrial Standardisation Law, Japan 193
industrialisation, and environmental degradation 156, 165–6, 217–18
information dissemination
by SAARC 89–90
by UNCTAD 65
on eco-labelling 55–6
inadequacy of 107, 120–1, 139, 144, 149–50, 220
inland waterways, South Asia 84–5
Intergovernmental Group on Trade Liberalisation (IGG), SAARC 90–1
international cooperation, trade and environment 88–91, 203–8
International Finance Corporation 176
International Institute for Sustainable Development (IISD) 209, 211, 212–15
International Monetary Fund (IMF) 214
International Pollution Deterrence Act (1991), US 173
International Standardisation Organisation (ISO) 35, 64, 65, 110, 150, 194
international trade *see* trade
intraregional trade, South Asia 88–91
investment, in environmental improvement 217–19
ITC 144, 153

Japan, environmental law 193–4
joint ventures 197–8
jute packaging 46, 62–3, 125, 145

Karnal technology 114
Karnataka Antibiotics and Pharmaceuticals (KAPL) 190–1

labelling
product 43
see also eco-labelling
land usage, South Asia 82–4
leather industry 97–103, 117–22, 136–7, 145, 151–2
in Bangladesh 97–103
banned use of PCPs 42, 120–2, 125, 177, 189
eco-labelling 119, 136–7
environment and trade linkages 117–22
footwear 99–100, 136–7
in India 42, 117–22, 125, 136–7, 189
in Nepal 145
in Pakistan 151–2
pollution 33–4, 101–3, 121, 151–2
life cycle *see* product life-cycle

'like product' rules 24–5
L'Occitane 206
Luigie Mettalurgie 197

Maldives 83–4, 85, 86, 87
marine environmental damage 85
market access
  and eco-labelling 27–8, 46–7, 58–9, 177
  and environmental policies 41–9, 218
  and PPM standards 25–9
  see also entry barriers; protectionism; trade restrictions
Marrakesh 124, 135
Max Havelaar Foundation 65, 206
microeconomic policies, for sustainable development 209–10
monitoring, environmental policies/standards 87, 114, 217, 219
Montreal Protocol 30, 87, 110, 123, 128, 179
  Global Environmental Facility 32
  Multilateral Fund 32, 173, 179
  ozone depletion 123, 128, 173
MST textile eco-labels 56–9, 132, 189
Multifibre Arrangement 64
multilateral environmental agreements (MEAs) 16, 30, 120, 126, 139, 172–3
  international commodity related environmental agreements (ICREAs) 36
  see also Montreal Protocol; Rio Declaration; World Trade Organization (WTO)
Multilateral Fund, Montreal Protocol 32, 173, 179
multinational corporations (MNCs), EST transfer 197
MUT textile eco-labels 56–9, 189

National Appliance Energy Conservation Act, US 42
National Conservation Strategy, Pakistan 149
National Leather Development Programme, India 122
Nepal 82–3, 87, 143–6
Netherlands Framework Agreement on Tropical Timber (NFATT) 27
Nippodenso, Japan 191–2
non-governmental organisations (NGOs)
  and EST transfer 197
  see also aid agencies
North American Free Trade Agreement (NAFTA) 153, 163, 210–12, 214

OECD
  Development Assistance Committee 203, 206
  environmental criteria inappropriate to developing countries 17–18, 34, 47, 55–6, 92
  regulatory regimes on cleaner technology 75
OECD countries 38, 125–7, 160, 176–7, 207, 218
  EST 175–6
  market for 'green' products 27–8, 33, 123–4, 143, 145–6, 149–50, 205–7
  relocation of polluting industries 26, 70–4
Oko-Tex Standard and certification 189
Organic Chocolate Manufacturers 206
ozone depletion 119, 123, 128, 172

packaging regulation 44–6, 48, 51–66
  compliance costs 45, 48, 127
  German Green Dot and Dual System (DSD) 51, 60–1, 151
  German Packaging Ordinance (1986) 45, 51, 59–63, 127, 131, 192
  IISD principles 214
  impact on developing countries 106, 119, 145, 159, 192
Pakistan 149–54
  eco-labelling 150–1

environmental degradation 83, 84, 85, 86
environmental impact assessment 87
leather industry 151–2
textile industry 150–1, 152
'eco-cotton' 150–1
trade–environment compatibility 149–54
paper industry *see* pulp and paper industry
pentachlorophenol (PCP) 42, 120–2, 125, 177, 189
pentane 129
pesticides
in Bangladesh 87
in India 114, 125, 130–1, 134–5, 193
Japanese regulation 193
in Pakistan 152
pharmaceutical industry, India 190–1
polluting industries 70–6
dyestuffs 133
leather industry 33–4, 101–3, 121, 151–2
share of world trade 70–1
pollution
'buyer pays principle' 16
'polluter pays principle' 16, 31
in South Asia 84–5, 86, 101–3
pollution control
costs 26
expenditure 175–6
harmonisation of 18–19
and industrial migration 26, 37, 69–78, 179–80
polychlorinated terphenyl (PCT) 190
poverty, and environment 15–16, 17, 53, 118
process and production method (PPM) standards 2, 4–5, 21–38
certification 2, 35–6
compliance costs 25–6, 28
and eco-labelling 33–5, 47
effect on international trade 23, 24–9, 31–6
'end of pipe' 113, 119, 174–5, 179
and 'green' trade preferences 27–8, 33
harmonisation 29–31
and product standards 22–4
see also environmental standards
process standards 21–2
product labelling 43
see also eco-labelling
product life-cycle, environmental regulation 44, 113, 178, 206–7
product standards 21, 22–3
see also environmental standards; process and production method (PPM) standards
product-specific policies 41–4
protectionism 17–19, 23, 64, 66, 117
see also entry barriers; market access; trade restrictions
Public Health Alliance, UK 163
pulp and paper industry 27, 43, 45–6, 55
PVC 44

quality assurance standards 23–4

Rainforest Alliance 206
recycling 43, 44–6, 48
German Dual System (DSD) 51, 60–1, 151
refrigeration 127–30, 191–2
relocation of production, and pollution control 26, 37, 69–76, 179–80
Rio Declaration 15, 37, 81, 203

Shriran Refrigerators 191
Singapore, Green Label system 47
South Asia
development and environmental policies 81–92
environmental degradation 82–6
environmental impact assessment 87
environmental protection 86–8, 92, 93–4 (*Appendices*)
impact of packaging measures 62–4
intraregional trade cooperation 88–91
see also Bangladesh; Bhutan; India; Nepal; Pakistan; Sri Lanka

South Asian Agreement on Regional Cooperation (SAARC) region
- competitiveness of polluting industries 71–4
- environmental cooperation 106
- harmonisation of standards 90–1, 92
- institutional changes needed 89–91
- relocation due to environmental standards 69–76

South Asian Cooperative Environment Programme (SACEP) 88
South Asian Preferential Trading Arrangement (SAPTA) 81–2, 91–2, 106
South Korea 176
Sri Lanka 155–60, 161–7
- Central Environment Agency 165
- eco-labelling 159
- effect of packaging regulation 159
- environment and trade linkage 155–60, 161–7
- environmental degradation 82–5, 156–8
- environmental impact assessment 87, 158, 165, 166
- environmental regulation enforcement 158, 165–6
- forestry 157
- hazardous substances 157–8, 165
- mineral exports 159
- sustainable development 155–6
- tourism 156, 157, 166
- trade liberalisation 156–9, 163–4, 165
- vehicle usage 157

standards *see* environmental standards; process and production method (PPM) standards; product standards
Sterlite Industries 197
Subros, India 191–2
subsidies, environmental 28–9, 30–1
Sustainable Agriculture, Food and Environment Alliance, UK 163
sustainable development 15, 16, 81, 203
- in Bangladesh 97–103
- in Bhutan 106–7
- IISD principles for 212–15
- in India 124–5
- microeconomic policies for 209–10
- role of trade 203–5
- in Sri Lanka 155–6
- and trade liberalisation 156–9, 204–5, 217–18

Taiwan 61, 176
taxes *see* economic instruments
tea production 2, 130–2
Technical Barriers to Trade (TBT) Agreement, GATT 24, 28
technology *see* environmentally sound technology (EST)
technology transfer 185–6
- of EST 10, 110–11, 113–114, 160, 171–81, 185–201; access by developing countries 160, 172, 176, 180–1, 195–6; by foreign direct investment 179–80, 196–7; by joint ventures 197–8; by patents 179; by purchase and licensing 177–9; 'clean technology Bank' concept 111; incentives for 196, 197–8, 199; in India 187–99

TELCO of India 189–90
testing facilities for environmental standards 122, 131, 136, 138
textile industry
- 'eco-cotton' 150–1, 189
- and eco-labelling 47, 51, 57–9, 63–4, 137–8
- in India 137–8, 188–9
- *see also* cotton industry

timber industry and forests 24, 27–8, 44, 206
- deforestation 82–3
- in India 17, 112–13, 131–2
- in Sri Lanka 157

Tokyo Agreement on Technical Barriers to Trade (1979) 24
tourism 86, 111
- in Sri Lanka 156, 157, 166

trade
  and environment 19–20, 155–60, 161–7, 203–8; in Canada 210–15; in South Asia 2–3, 81–92
  exports of polluting industries 70–4
  impact of packaging measures 62–4
  South Asian intraregional 88–91
  see also trade–environment mutual compatibility
trade liberalisation
  in developing countries 2–3, 16, 90–2, 162, 163–4
  and environmental standards 3, 123
  in India 16, 111–13, 120
  in Sri Lanka 156–9, 163–4, 165
  and sustainable development 156–9, 204–5, 217
trade restrictions
  and environmental concerns 15, 23, 31–2, 37, 118–20, 173
  see also entry barriers; market access; protectionism
trade–environment mutual compatibility 209–15, 217–20
  in Bhutan 105–7
  in India 109–14, 117–20, 123–39
  in Nepal 143–6
  in Pakistan 149–54
transaction costs 45, 48, 127
Turkey 47
TWIN/TWIN TRADING 206

UNCTAD 65, 144
  Working Party on Interrelationship between Investment and Technology Transfer 187
  Working Party on Trade, Environment and Development 34
United Kingdom 163
United Nations Development Programme (UNDP) 88, 122, 144, 153
United Nations Environment Programme (UNEP) 88, 153
United States 42, 173–4, 191
Uruguay Round, GATT 1, 4, 134, 135, 161–2, 163–4, 218
  Technical Barriers to Trade (TBT) Agreement 24, 28

vehicle exhaust emissions 189–90

waste
  trade in 60–1, 204–5
  waste management 43, 45–6, 59–63
  see also packaging; recycling
Waste Avoidance and Management Act (1986) 59–63
WIDER report on Indo–Sri Lanka Economic Cooperation 160
wildlife
  endangered species 136
  habitat 83–4, 139
World Bank 176, 214
World Health Organisation 88, 191
World Trade Organisation (WTO) 75, 161–2, 163, 214
  and environmental subsidies 28–9, 30–1
  involvement in environmental concerns 15, 64, 66, 110, 122, 163–4
  'like product' rules 24–5
  Tokyo Agreement on Technical Barriers to Trade (1979) 24
  see also Uruguay Round, GATT